Martin Carrier, Rebecca Mertens, Carsten Reinhardt (eds.)
Narratives and Comparisons

BiUP General

Martin Carrier, born in 1955, teaches philosophy of science at Bielefeld University. His research is directed at methodological characteristics of science in the context of practice, that is, research targeted at economically or politically relevant fields.

Rebecca Mertens (PhD), born in 1984, is a postdoctoral researcher in the history and philosophy of science at Bielefeld University. She works on the role of analogies, models and forms of comparison in the history of molecular genetics and is a member of the collaborative research program »Practices of Comparison: Ordering and Changing the World«.

Carsten Reinhardt, born in 1966, teaches history of science at Bielefeld University. His research fields are the history of twentieth-century physical sciences, methods development, and expertise.

Martin Carrier, Rebecca Mertens, Carsten Reinhardt (eds.)

Narratives and Comparisons

Adversaries or Allies in Understanding Science?

BIELEFELD UNIVERSITY PRESS

The volume has grown out of the workshop »Practices of Comparing and Narrating in the Sciences« within the collaborative research center on Practices of Comparison (SFB 1288, Praktiken des Vergleichens).
The workshop took place at Bielefeld University in April 2018 and was organized by Veronika Hofer.

Bibliographic information published by the Deutsche Nationalbibliothek
The Deutsche Nationalbibliothek lists this publication in the Deutsche Nationalbibliografie; detailed bibliographic data are available in the Internet at http://dnb.d-nb.de

An Imprint of transcript Verlag

http://www.bielefeld-university-press.de

Cover layout: Maria Arndt, Bielefeld

Print-ISBN 978-3-8376-5415-8
PDF-ISBN 978-3-8394-5415-2
https://doi.org/10.14361/9783839454152

Contents

Introduction
Narratives and Comparisons: Adversaries or Allies in Understanding Science?

Martin Carrier, Rebecca Mertens, Carsten Reinhardt

1. Practices of Comparing and Narrating in the Sciences and Humanities

In this volume, we aim to explore the ways in which comparing is related to other social and epistemic activities in knowledge generation processes in the sciences, humanities and the arts.[1] We particularly emphasize the relationship between comparing and narrating in epistemic practice. One of our central goals is to clarify the potential of narratives for drawing comparisons, that is, the way in which narratives can enable, support or hinder the practice of comparing. Furthermore, this volume seeks to locate narrating and comparing within the conceptual and methodological, material, and discursive practices that are involved in knowledge generation processes. The focus on practice also means for us to look more closely at the relations between those activities and processes which form the conditions for using both comparisons and narratives successfully in the production of knowledge. We do this

1   The following chapters have been composed in the framework of the Collaborative Research Center SFB 1288 "Practices of Comparing. Changing and Ordering the World", Bielefeld University, Germany: "Introduction," "Historical Narrative versus Comparative Description? Genre and Knowledge in Alexander von Humboldt's Personal Narrative," "Narrating and comparing in the organization of research projects," "Seeing, Comparing, Narrating, Making–of the Middle-Ages in the Early History of Art," "Narrating Art History: Practices of Comparing in Exhibitions and Written Surveys with regard to documenta I." The work has been funded by the German Research Foundation (DFG), subproject Co4: Vergleichshindernisse in den Naturwissenschaften und ihre Überwindung. Das Beispiel der Molekulargenetik. We thank Gina Maria Klein for her support in finalizing the manuscript.

by taking into account related activities, such as measuring and classifying, modeling, and establishing norms and categories, as well as the organization and popularization of knowledge. In particular, we discuss and hopefully contribute to dissolving the often assumed opposition between the role of narratives in scientific explanation, on the one hand, and in understanding, on the other. We propose that narrative practice is closely linked to, if not even part of, what many scientific explanations achieve. We thus pay close attention to the explanatory role and potential of narratives as used in the natural sciences and the humanities, among others in historiography, anthropology and paleontology, as well as physics, biology and chemistry. In line with Kaiser et al. (2014) and Glennan (2010),[2] we see a close connection between the use of narratives in unfolding historical events, on the one hand, and as explanatory tools in the sciences, on the other.

It does not come as a surprise that we emphasize comparing as a crucial part in practicing the sciences and humanities. Comparing is such a powerful tool in the production of knowledge that it even became part of the very names of disciplines: comparative anatomy, comparative law, and comparative literature studies. In historiography, comparing became one of the methodological cornerstones of modern social and cultural history.[3] In general, and not related to just the sciences and the humanities, the importance and ubiquity of comparison in the establishment of social order and norms, as well as a means of grasping the unknown in (inter-)cultural encounters, has recently been emphasized by Angelika Epple and Walter Erhart. Drawing comparisons is understood here as an activity which is always situated within particular cultural practices, enforcing certain ways and ends of comparing and suppressing others, thereby strongly shaping social discourses and structures.[4] One of the best known postulates about the constitutive role of certain

---

2   Kaiser, Marie, and Daniel Plenge, "Introduction: Points of Contact between Biology and History." In *Explanation in the Special Sciences. The Case of Biology and History*, edited by M.I. Kaiser, O.R. Scholz, D. Plenge, and A. Hüttemann. Amsterdam: Springer, 2013. See as well Glennan, Stuart, "Mechanisms, Causes, and the Layered Model of the World," *Philosophy and Phenomenological Research* 81(2), 2010: 362-381.

3   Welskopp, Thomas, "Comparative History," *European History Online (EGO)*, published by the Institute of European History (IEG), Mainz 2010-12-03. URL: http://www.ieg-ego.eu/welskoppt-2010-en URN: urn:nbn:de:0159-20100921414 [2019/06/24].

4   Epple, Angelika, and Walter Erhart, "Die Welt beobachten. Praktiken des Vergleichens." In *Die Welt beobachten. Praktiken des Vergleichens*, edited by A. Epple and W. Erhart, Frankfurt a.M.: Campus, 2015, 7f.

forms of comparison for modern culture has been introduced by Michel Foucault in *The Order of Things*, where he identifies difference making as one of the most important epistemic activities in the history of modern Western thought, while claiming that "difference making" partially replaces analogical thinking and similarity-driven strategies. In keeping with such approaches, we underline the epistemic significance of comparing. Drawing comparisons involves at least two *comparata* and a *tertium comparationis*, the latter defining the perspective of the comparison in question.[5] Thus, over the last 50 years, comparing has been established as a well developed, and well connected, epistemic activity in the sciences and humanities, placed alongside and demarcated from other methodologies.

While science and cultural studies have by now carved out a widely recognized role for comparing in the sciences and humanities, the same cannot be said about narrating. Although recent studies emphasizing the crucial importance of metaphors and analogies in research[6] can in parts be traced back to classics in the field, such as the work of Ludwik Fleck,[7] the study of narration has entered science studies at a broader scale only very recently.[8] Partially, this might be due to the preconceived notion that telling a story is, at best, part of communicating results, and certainly not crucial in producing them. This view is challenged at several levels in the contributions to this volume. Moreover, even well founded claims about establishing the central role of narratives

---

5   Epple, Angelika, "Ein praxeologischer Zugang zur Geschichte der Globalisierung/en." In *Die Welt beobachten. Praktiken des Vergleichens*, edited by A. Epple and W. Erhart, Frankfurt a.M.: Campus, 2015, 163. See as well Sass, Hartmut von, "Vergleiche verstehen. Einleitende Vorwegnahmen." In *Hermeneutik des Vergleichens. Strukturen, Anwendungen und Grenzen komparativer Verfahren*, edited by A. Mautz and H.v. Sass, Würzburg: Königshausen & Neumann, 2011, 27.

6   Mertens, Rebecca, *The Construction of Analogy-Based Research Programs. The Lock-and-Key Analogy in 20th Century Biochemistry*, Bielefeld: transcript Verlag, 2019. Frigg, Roman and Stephan Hartmann, "Models in Science," *The Stanford Encyclopedia of Philosophy* (Summer 2018 Edition), edited by Edward N. Zalta, https://plato.stanford.edu/cgi-bin/encyclopedia/archinfo.cgi?entry=models-science. Morgan, Mary S., *The World in The Model. How Economists Work and Think*, Cambridge (UK): Cambridge University Press, 2012.

7   Fleck, Ludwik, *Genesis and Development of a Scientific Fact*, Chicago: University of Chicago Press, 1979.

8   Morgan, Mary S., and Norton Wise, "Narrative Science and Narrative Knowing. Introduction to Special Issue on Narrative Science," *Studies in History and Philosophy of Science* 62, 2017: 1-5.

in research have been disputed. An example is Hayden White's *Metahistory*,[9] which was criticized for its rather narrow methodological approach and the fact that it kept being restricted to a single field. The contributors to this volume both increase the number of methodological access points and widen the range of study fields. As a result, the role of narratives in a wide spectrum of scientific and humanistic fields is brought into view.

Arguably, the reasons for changing the status of narrating in the sciences and humanities for the better have emerged from two recent developments. The first has its origins in the field of science studies proper. Since the beginning of social studies of science, the influence of the social context in one way or another on scientific practice and content has been a defining feature, giving rise to new subfields, such as social epistemology. Among those, the claim that social, cultural and political validity are crucial sources of scientific authority has recently become a staple of studies in political epistemology.[10] The key question that follows from this view for us is how do social acceptance, societal validity and epistemic authority act back on research practices, and in what circumstances do narratives play a role in this? Recently, Safia Azzouni and Stefan Böschen, in their introduction to a volume linking narration and scientific validity, have pointed to some general characteristics in this regard.[11] In their view, scientific and social actors create narrations, or scenarios, of how the features of the scientific problem at hand are related to each other. These scenarios often compete with or even exclude each other, and this adds to the difficulty for achieving scientific and societal consensus. The debates on climate change are a case in point where questions of evidence come to the fore: evidence by whom, for what aim, and to what outcome? Narratives play a central role in such debates, and justly have become a key entry point of science studies for analyzing various constellations of science in society.

---

9   White, Hayden, *Metahistory. The Historical Imagination in Nineteenth Century Europe*, Baltimore: Johns Hopkins Univ. Press, 1973.

10   Straßheim, Holger, "Politics and Policy Expertise." In *Handbook of Critical Policy Studies*, edited by F. Fisher et al., Cheltenham UK / Northampton MA: Edward Elgar, 2015, 319-340.

11   Azzouni, Safia, and Stefan Böschen, "Erzählung und Geltung. Ein problemorientierter Ausgangspunkt und viele Fragen." In *Erzählung und Geltung. Wissenschaft zwischen Autorschaft und Autorität*, edited by S. Azzouni, S. Böschen, and C. Reinhardt, Weilerswist: Velbrück Wissenschaft, 2015, 9-31.

This fruitful development in science studies has been strengthened by a movement in narrative studies that has broadened the concept of narrative and at the same time has introduced additional fields of analysis. This is the second development that we wish to highlight here. More and more it has become accepted in literature studies to include non-fictional texts in studying narration. This has not been traditionally the case, though. In the last decade or so, however, the narrative border between non-fiction and fiction has been blurred by highlighting their structural similarities. While the range for studying narratives has been widened, so has the understanding of narrative. In this view, narratives are not just stories, and narrating is much more than mere story-telling even though it includes the latter. Here, narratives are understood as higher-order structures of how stories are related to each other. They constitute patterns of story-telling in both their temporal and structural dimensions. While single stories are based on a certain timeline or stage-setting, and establish individual links between the events told and the agents described, narratives in addition often implement general temporal and configurational patterns of what counts as a gripping timeline and a convincing plot. "Good beats Evil," or "Success after severe obstacles have been overcome" are examples.[12] It is important to note that narratives contain both temporal and structural or configurational dimensions in constituting such higher order patterns of both timeline and stage-setting (see below sect. 4).

Thus, the study of narratives points to key features of reasoning in the sciences and humanities. Narratives contribute to the analysis of causality and contingency in offering patterns of how the entities in question are linked and in outlining what kinds of processes play a role in their evolution. It thus may very well be that narratives support epistemic practices of drawing comparisons in the sciences and the humanities. Narratives sometimes enable or facilitate comparisons and explanations.

2.   Comparison and Narrative as Methodological Tools

In the following sections, we stress this productive role of narratives for drawing comparisons by addressing features of scientific practice. Drawing comparisons is an important methodological tool for producing order in conceptual respect and for coping with new and unaccustomed objects and phenom-

---

12   Azzouni and Böschen, "Erzählung und Geltung," 17.

ena. What we do in creating order is assimilating what is unknown in part to what we are familiar with. Thus, we isolate and delimit the novel and un-expected elements. This is achieved by invoking relations of similarity so that hopefully merely a few elements are left that do not fit established categories. Accordingly, the method of drawing comparisons serves to anchor new items at familiar piers and helps us navigate through uncharted waters of bewilder-ment and surprise.

However, the handy tool of comparing does not always work well in gen-erating transparency and clarity. Drawing comparisons is jeopardized by ob-stacles and barriers. Sometimes it is difficult to establish a common ground or a shared yardstick that could provide a standard of similarity against which differences could be compared. For instance, it is difficult to assess whether a sufficient amount of significant features is shared between the concepts of chemical element before and after the Chemical Revolution. Prior to the Rev-olution, chemical elements were thought to be abstract bearers of chemical properties such as combustibility or acidity. Their abstract nature was sup-posed to imply that elements are not material in themselves. Elements are no substances and cannot be isolated in chemical analysis for this reason. They rather explain properties of material substances. If they were material themselves, we would run into a circularity. After the Chemical Revolution, elements were conceived as end-points of chemical analysis and are thus defi-nitely to be found in the laboratory. The explanatory and the operational con-cepts of chemical element are widely disparate and threatened by *non-com-mensurability*, i.e., the lack of significant shared properties.

In a different vein, Kuhnian *semantic incommensurability* is produced by a cross-classification of similarity classes. For instance, the early notion of virus broke the confines of the then-contemporary cluster of properties assigned to bacteria and toxins, respectively. Viruses are contagious and reproduced in or-ganisms, which showed them to be of the same kind as bacteria, but they pass through filters that withhold bacteria. Therefore, viruses could not be cells and looked rather like toxins. As a result, applying the standard procedures for categorizing biological entities generated conflicting judgments. The newly discovered entities transcended existing conceptual boundaries and defied comparative analysis for this reason.

How can such obstacles to comparison be overcome? One of the options is to introduce intermediate stages that combine features from both ends of the conceptual spectrum or to invoke transitory steps that gradually lead from one end to the other. A bridge notion of chemical element introduced the idea of active substances. Elements are recognized as material substances, to be sure, but they are considered capable of imposing their properties on other, more passive substances. Regarding viruses, early researchers conceived them as cellular fluids. They were supposed to be part of cells and thus reproduced within and together with cells, but they were thought to migrate among host cells and could move independently of them. Such intermediate conceptual states are often produced by appealing to analogies and metaphors. This volume thus aims to clarify the role narratives could play in the endeavor of establishing relations of similarity and making comparisons possible.

3.   Narratives in the Temporal Sense and their Roles in Comparison and Explanation

As we said before, providing a narrative means outlining a plot and a stage setting. The primary understanding of story-telling is to produce a time series of events. Two disparate stages are connected by a temporal sequence of intermediate states, and narratives are patterns for how to trace such time evolution in the phenomena studied. In this vein, each of the relevant steps involves only a small-scale change, but adding up such steps may serve to connect seemingly contrasting states of affairs. Consequently, narratives may enable or facilitate comparisons. Likewise, narratives provide explanations in virtue of the ties that bind subsequent states together. Such states may be bound by causation or by biological evolution, and in virtue of such connections the common ground between apparently unrelated or incomparable stages is revealed.

For instance, sun-like stars, red giants, and white dwarfs look utterly dissimilar in their characteristics, but they are easily comparable once they are recognized as phases of stellar development. When sun-like stars have exhausted their hydrogen resources in the core, thermonuclear fusion moves outward to the shell surrounding the core. Their size is thereby greatly expanded while the surface temperature is reduced because in virtue of their vastness the radiation emitted is distributed over an enormous surface. When nuclear fuel is used up eventually, the star collapses into a tiny and dense stellar remnant of faint luminosity. This story gives a causal explanation of stellar evolution and shows that these apparently disparate phenomena can be compared by placing them on a causal and temporal scale.

Biological evolution is another stronghold of narratives. The Darwinian mechanism of inheritance with variation and the selection of organisms by environmental conditions means to account for the present state of affairs by invoking past constellations. Such historical explanations are sometimes contrasted with rational explanations in that seemingly less than optimal results are traced back to different, earlier conditions. Think of the Brazilian variant of the green sea turtle (*Chelonia mydas*) which reproduces only under environmental conditions that are different from their usual habitat. This variant migrates over a thousand miles to the mid-Atlantic Ascension Island for laying its eggs, although suitable conditions also exist much closer to their home turf. The assumption is that this seemingly bizarre behavior evolved at a time when the Atlantic Ocean was much smaller. The reproductive pattern was sensible at its inception but grew increasingly peculiar by continental drift.[13] The evolutionary narrative restores biological sense to an apparently odd behavior. This is achieved by positing an initial state and then tracing intermediate steps to the present condition. The explanatory power of this story is based on bridging this initial state and the observed situation with a sequence of transitional states. In this way, earlier and later stages are made comparable.

A similar case can be made for conceptual development in the history of science. Oxygen, as conceived by Lavoisier, has barely anything in common with the modern notion. Oxygen in its present-day understanding does not underlie the nature of acidity, nor is it bound to the matter of heat that it gives off in forming a compound in combustion, which was both the case in Lavoisier's conceptual frame. But in tracking gradual conceptual changes in history, we can realize that earlier and later stages are connected. They are not connected by one thread running through all stages, but by a variety of shared features changing in each step. That is, historical sequences may be tied together by narratives that establish Wittgensteinian family resemblance among the stages.[14]

In sum, narratives may tell a causal story or an evolutionary history in virtue of which we are able to grasp the relationship between two seemingly distinct states of affairs. The two states become comparable in that a gradual transition leads from one to the other. In this vein, time sequences have been claimed to provide explanatory resources in fields traditionally considered to

13   Gould, Stephen Jay, *The Panda's Thumb*, New York: Norton, 1980, 30-33.
14   Chang, Hasok, "The Persistence of Epistemic Objects Through Scientific Change," *Erkenntnis* 75, 2011: 413-429.

be governed by universal and eternal laws of nature. In addition to realizing the importance of temporal developments in fields like astrophysics (e.g., stellar evolution), philosophers and historians of science have claimed that a marked methodological shift is underway. This shift involves the growing importance of computer simulations. The primary mode of explanation is said to be no longer to subsume phenomena under higher-order laws of nature. Rather, phenomena are understood by pursuing how more elementary objects build up or are grown into more complex ones. As Norton Wise claims, simulation techniques trace the development of an object through changing conditions and elucidate in this way the features of this object. Such techniques provide a story as to how this object has come about and explain its properties by following its development through a series of changes. As a result, there is no opposition between narrative and explanation. On the contrary, computer simulation widens the scope of explanation by bringing individual variation into its purview. Real snowflakes, in contrast to their idealized image, are non-symmetrical and variable, and these features can be accounted for by simulating the growth of snowflakes under a variety of conditions. Such diversity can be expounded best by tracing many individual trajectories under different initial and boundary conditions. This is what computer simulations accomplish, and this is why they usher in a new narrating mode of explanation.[15]

Turning to evolutionary biology, Wise's claim that variability can best be explained by narrative explanations is confirmed from a different angle by John Beatty. He argues that giving explanations by following historical lines is of chief importance in evolutionary biology and claims that an essential element of valuable narrative explanations is contingency. More specifically, narrative explanations are indispensable when we are faced with a branching-tree scenario in which alternative pathways open up at various junctures. The path picked at such "turning points" makes a difference for the future course. Put conversely, a particular outcome can only be explained by tracing the choices at the turning points and pursuing the actual pathway through the branching tree of non-actualized possibilities. As Beatty emphasizes, it

---

15   Wise, Norton, "Introduction: Dynamics all the Way Up." In *Growing Explanations. Historical Perspectives on Recent Science*, edited by N. Wise, Durham: Duke University Press, 2004, 1-20. Wise, Norton, "Science as (Historical) Narrative," *Erkenntnis* 75, 2011: 349-376.

is the existence of such turning points that makes narratives essential and a story worth being told.

Contingency is a critical factor of a persuasive narrative. We could develop a phenomenon governed by deterministic laws (such as the motion of planets) into a series of events (giving the positions of the planets at different times). But this would be entirely pointless. Valuable narratives need to contain contingency or turning points. Beatty distinguishes between contingency *per se*, such that an event was not bound to occur and could have come to be otherwise, and contingency upon previous events, in the sense that a subsequent event depends on earlier events for happening. The turning points of an interesting narrative need to be contingent in both senses. If a sequence of organismic states is to be accounted for by an interesting evolutionary narrative, two conditions need to be realized: First, later states of this sequence need to be contingent on earlier states such that the later states would not have evolved if the earlier states had been different. Second, early states need to be contingent *per se*. It might well have been the case that earlier stages of the organism might have gone extinct. This makes the survival of the earlier stage a turning point and transforms the whole episode into a worthwhile narrative.[16]

Staying with evolutionary explanation, one of the traditional complaints about such a narrating mode of explanation concerns the arbitrary nature of the historical steps assumed. In a celebrated contribution, Stephen J. Gould and Richard C. Lewontin have castigated the carelessness of biologists in thinking up evolutionary stories for accounting for an observed state. Countless unsupported hypotheses are produced for explaining why a given trait of an organism is useful for survival. For instance, it was reported that male bluebirds are more jealous before mating than after completed copulation. This was supported by registering the number and fierceness of attacks of bluebirds on dummy birds close to their nest. The explanation lies right at hand: when mating is accomplished the male bluebird can be sure that his genes are in the eggs. Jealousy would be futile; the optimum evolutionary strategy is to let the competitor exhaust his forces in vain. However, the result of the study could not be reproduced in a follow-up experiment. In this second experiment, jealousy was observed to be weak all the time and no changes

---

16   Beatty, John, "What are Narratives Good for?," *Studies in History and Philosophy of Biological and Biomedical Sciences* 58, 2016: 22-40.

were recorded. Biologists were quick to produce alternative hypotheses. Obviously, female bluebirds were available abundantly and the best strategy for a male bluebird was simply to leave an unfaithful female rather than running the risk of entering into a fight with a competitor. Gould and Lewontin are highly critical of such "pan-adaptationist" strategies which invent selective advantages copiously and arbitrarily and replace them without much ado in the rare event of counterevidence arising. "Just-so stories" fail to explain.[17]

Being mindful of arbitrariness and insisting on independent empirical support is certainly a recommendation worth being heeded. The initial and intermediate states introduced or appealed to an acceptable explanatory narrative should be confirmed by observational evidence. However, this important caveat should not be exaggerated either. Sometimes evolutionary explanations are of the "how possible" variety. That is, how is it conceivable that a certain complex organ developed by variation and selection and that a certain species grew out of a prima-facie quite distinct ancestor species? In such cases, evolutionary trajectories that are merely possible serve the purpose of dispelling the mystery. It is true, it would be better even in such cases if supporting evidence were offered. But even without such assistance, the main explanatory purpose is served by the evolutionary narrative, namely, showing that an evolutionary pathway is available in the first place that connects the two states in question.

This sketch of some of the key positions on offer is supposed to show that narratives in the temporal sense are able to establish comparability and explain features in the physical and biological world. Reconstructing how an outcome has been produced and how it has grown out of preceding states can illuminate key features of this outcome. The sequence of subsequent states exhibits how these states are related to each other and how the causal factors are involved in the production and variability of its features. In this vein, narrative accounts may yield explanations of the particulars of the phenomena at hand that are inaccessible to more abstract and universalist approaches,

---

17   Gould, Stephen J., and Richard C. Lewontin, "The Spandrels of San Marco and the Panglossian Paradigm. A Critique of the Adaptationist Programme." In *Conceptual Issues in Evolutionary Biology. An Anthology*, edited by E. Sober, Cambridge MA: MIT Press, 1984, 252-270 [1978]. Recently, Kaiser and Plenge have challenged the explanatory value of "just-so-stories" in the context of historical narratives, building up on Gould's and Lewontin's canonical argument. See Kaiser, Marie, and Plenge, Daniel, "Introduction: Points of Contact between Biology and History." In *Explanation in the Special Sciences: The Case of Biology and History*, 1-23, on p. 9.

and giving such explanations invokes and presupposes comparisons by intro-ducing transitory stages that mediate between posited initial states and the observed outcome.

4.   Narratives in the Coherentist Sense and their Roles in Comparison and Explanation

A non-temporal understanding of narratives has been developed in the past decades and has been found increasingly appealing more recently. In this ap-proach, the chief role of narratives is to establish a colligation of phenomena or to create coherence. An early proponent of this view is Stephan Hartmann who studied the relationship between a formal theory or theory-based equa-tions, on the one hand, and "stories" (in a non-temporal sense to be clar-ified), on the other. Hartmann's account features the relationship between Quantum Chromodynamics (QCD), the theory of strong interaction that is effective in the atomic nucleus, and the properties of hadrons, i.e., particles made of three quarks. Since the formal equations are insufficient for reach-ing concrete explanations and predictions, models have been developed that take special initial and boundary conditions into account and achieve such explanations for restricted features of the particles under consideration. As Hartmann portrays the situation, two such models are in play that empha-size different features and neglect others. In order to justify these one-sided approaches, stories are told, respectively, why the selected features are im-portant and the neglected ones are insignificant. Models thrive on the story told.

The models mostly provide causal mechanisms. These mechanisms are neither part of the theory nor deducible from it, but they are inspired by its formalism. The story complements the formalism and fits the model into a larger framework in a non-deductive way. In other words, a model is an in-terpreted formalism enriched by a story. For instance, one of the competing models, the so-called MIT-Bag Model, owes its popularity to the convincing story going along with it. The model concerned the failure to observe free quarks, and the attached narrative sketched a causal process by which a quark created an interaction-free space around it. This means that the quark was shielded from outside intervention and could not be exposed. Although ref-erence to a causal mechanism contains an implicit time element, this was not actually used in the narrative. By adding a causal account to the theory, the narrative provides an understanding of the physical processes in play.[18]

The non-temporal understanding of narrative explanations has chiefly been advanced by Mary S. Morgan. In her "configurational" account, a narrative explanation is characterized by its ability to order materials by temporal, spatial, theoretical or conceptual relations. Such an explanation is achieved by binding events together (colligation) or by contrasting them (juxtaposition). This means that sets of relevant elements are picked out and contrasted with less than relevant items. Morgan insists that this production of order is the crucial achievement of narratives, not the use of temporal relations for this purpose.

Narratives are distinguished from models by being concrete and particular. However, they use theoretical categories and conceptual elements for sorting concrete items and this is why they can be generalized to different cases. Narratives are thus generic by nature, and thanks to their conceptual structure they can be transferred to other fields. One of Morgan's examples is a study on the "street corner society" in a Boston slum area, conducted by William Foote Whyte in 1943. The study introduced a variety of groups and examined their relations: for instance, ill-educated and impoverished young men in contrast to better educated and better-off young college men, their relations with racketeers in contrast to those with the police. Slums had been considered socially amorphous venues before, while in the light of the study they appeared as organized social bodies with a social hierarchy in place and a recognized system of social obligations.

The study employed group behavior, leadership, and community as concepts for colligating and juxtaposing behavior and thus for creating coherence. The use of such general concepts made it possible to transfer the discovered phenomena to other communities. The study effectively coined the generic term "slum society." Such narrative explanations work by showing that events are related in a certain way so that they become significant and intelligible. Morgan compares such narrative explanations to mosaics, jigsaws or collages, in which the parts acquire their significance through their relation

---

18   Hartmann, Stephan, "Models and Stories in Hadron Physics." In *Models as Mediators. Perspectives on Natural and Social Sciences*, edited by M.S. Morgan and M. Morrison, Cambridge: Cambridge University Press, 1999, 326-346.

to other elements. They lack a story line, but they resemble documentaries which also analyze and show how the elements involved hang together.[19]

Narratives in this configurational or coherentist sense are characterized by turning to the details and particulars. They may be invoked to provide causal mechanisms or to establish order by sorting items into categories. Time order does not play a significant role in either variant. Rather, explanations are supplied and order is created by attending to concrete events and phenomena by using general concepts. In this way, narratives are usually related to theories, but the claims they entail are independent of theories. Yet, thanks to drawing on theoretical categories, narratives are generalizable and thus pave the way toward giving explanations.

The relationship of such narratives to comparison leaps to the eye. Morgan's emphasis on creating clusters of similarity and dissimilarity obviously thrives on drawing comparisons. Hartmann's stress on causal stories involves comparisons between the envisaged mechanisms and their theoretical embedding. In both accounts drawing comparisons is a key activity in providing narrative explanations.

In sum, narratives in the temporal and the configurational sense establish relations among entities in question and thereby enable comparisons. Such comparisons are in their turn an important element in providing explanations. On the whole, narratives are of instrumental significance in many epistemic practices in the sciences.

5.   Overview of the Volume

Part I of the volume centers on the general, conceptual dimensions of the interplay of narrating and comparing in the sciences. Norton Wise opens the volume with the chapter "Does Narrative Matter? Engendering Belief in Electromagnetic Theory." He takes up and develops further the configurational account of narratives. In agreement with Hartmann, Wise stresses that the familiar criteria for judging theories, such as empirical adequacy or mathematical consistency, are often not sufficient for singling out a particular account as superior. What is needed in addition for supplying credibility is a narrative.

---

19   Morgan, Mary S., "Narrative Ordering and Explanation," *Studies in History and Philosophy of Science* 62, 2017: 86-97.

Wise's examples are taken from the nineteenth-century electrodynamic theories of Maxwell and Weber. Both crafted an avowedly fictional framework which, in Maxwell's case, assigned a place to lines of force and, in Weber's case, to action at a distance. These frameworks established relations, if imaginary, among the various elements involved. Both were anchored in features existing in the real world, such as flow analogies in the case of Maxwell or operations in the case of Weber, and introduced fictional features into these familiar settings. In Maxwell, conceptual metaphors such as "lines of force" served to link up events in the everyday world to the fictional features.

Hybrid constructions of this sort constitute narratives in Wise's understanding. Following Sarah Johnston's account of Greek mythology, Wise uses her concept of a "story world" for denoting this combination of an existing and familiar setting with interspersed imaginary components. Fictitious entities are embedded in a realistic framework, and it is this connection to the real world that conveys plausibility to the fictitious parts. Accordingly, the distinctive feature of a narrative is to bring characteristics of different kinds together in one conceptual scheme. While narratives may use time-order to achieve such colligation, they need not. Rather, narratives serve to bind various aspects and details together and introduce contingency, possibility, probability, and alternatives. Narratives produce coherence by distinguishing such constellations from alternative ones. As a result, narratives emerge from drawing comparisons and they lead to comparisons.[20]

In contrast to Wise, Christine Peters takes up the temporal view of narratives in her chapter "Historical Narrative versus Comparative Description? Genre and Knowledge in Alexander von Humboldt's *Personal Narrative*." As she argues, depending on how narratives are conceived they can support or undermine the explanatory role of narratives. The latter possibility emerges if narratives are viewed as autobiographical stories. A "historical narrative" conjoins subsequent states of experience of an observer. The link between different such states is produced by the chronological order in which they were registered. With regard to the activity of comparing this means that similar states or comparable processes may be separated by a time lag; they enter the mind under different circumstances. Narratives based in this way on personal history can impede significant comparison and fail to arrive at substantial explanation.

---

20   See Morgan and Wise, "Narrative Science and Narrative Knowing."

Humboldt prefers a different narrative style that does not follow the traveler's experiences but rather renders the relations among the phenomena observed. Such "comparative description" is achieved by emancipating oneself from the accidental features of sequences of impressions supplied by the senses and by focusing instead on the relations among different observations. For instance, when Alexander von Humboldt managed to compare the changes in wild plants upon moving northward with such changes when moving upward to higher altitudes, he abstracted from the contingent circumstances of the relevant observations. In order to draw such comparisons, Humboldt needed to connect what was not linked by an uninterrupted temporal flow of observations. Rather, in his comparative description of volcanoes, Humboldt highlighted geological relations and causal connections. Description emphasizes the sequence of natural events.

This outline shows that Peters draws on the temporal understanding of narratives and locates the explanatory power of such narratives in their invocation of objective relations among natural events. Uncovering causal chains is an appropriate basis for a narrative explanation. The relevant time sequence is shifted away from the observer and toward the phenomena (see section 3 above). Both historical narratives and comparative descriptions are based on drawing comparisons. But only the latter are able to establish relations of similarity that pertain to the course of nature, and thus only the latter are suited to giving explanations.

Hans-Jörg Rheinberger deals with the "Narrative Order of Experimentation." The tradition of letting the objects of study tell their own stories has accompanied the sciences from their inception. Rheinberger argues that the extended process of experimentation can indeed itself be regarded as a form of narration. Furthermore, he distinguishes three levels on which such narratives can be accounted for historiographically. One level is experimental systems. Such systems stimulate the production of histories of exploration or micro-histories. The narratives created at this level are case-studies. However, second, such micro-histories stand for some more general state of affairs and therefore need to be embedded. Moving on to the temporal meso-range of a century rather than a decade brings experimental cultures into view. In-vitro experimentation is an example of an experimental culture. Taking such cultures as the object of a narrative upholds the emphasis on practice but transcends particular experimental conditions and laboratories. Such cultures are more fine-grained than disciplines and exhibit a focus on practice. Specific narratives can be told by regarding experimental cultures as the engine of innovation and by trying to generalize their impact to analogous cases.

A third level (in addition to experimental systems and experimental cultures) is represented by the macro-histories that can be unfolded by the transition to scientific concepts. For instance, around 1800, the concept of heredity changed its meaning profoundly in that emphasis was now placed on organisms as carriers of intergenerational property transmission. This understanding became encapsulated in later decades in the notion of the gene. This notion also underlies twentieth-century grand narratives such as the "geneticization of society." All in all, experimental practices in the sciences are intertwined with activities of narrating, and the latter can shed light on what scientific scrutiny and scientific change is all about.

Part II of the volume deals with the social, economic and political conditions of research practice, particularly with the role of comparing and narrating in research organization and popularization.

Oliver Hochadel explores the relationship between comparing and narrating in research on Neanderthals at the interface of archeology, paleontology and paleofiction. In the course of the prominent archeological finds in the Shanidar cave in Northern Iraq in the early 1950s, the image of the Neanderthals was about to be transformed from "beast to brother"—an image so powerful to lastingly influence our view of prehistoric life until the present day. Hochadel retraces the narratives that led to the rehabilitation of the Neanderthals, depicting them as social individuals with human-like behavior. He analyzes the work of the former anthropologist Ralph Solecki who was among the first researchers and popular science writers, portraying the Neanderthals from the Shanidar cave as compassionate and almost human individuals. In his analysis of Solecki's influential book "*Shanidar. The First Flower People,*"[21] Hochadel shows that the depiction of the Neanderthal rituals resulted from Solecki's "double field-work," involving the interpretation of archeological findings, on the one hand, and the observation of Kurdish life in contemporary Northern Iraq, on the other. As a result, a continuum between "the deep past and the present" and a convincing narrative of the relationship between Homo sapiens and Neanderthals was established. In the following decades, prehistorians and novelists would depict the Neanderthals as emotional and even social individuals who deeply cared for their community. Especially Solecki's depiction of the burial of one of the Shanidar Ne-

---

21   Solecki, Ralph S., *Shanidar, the First Flower People*, New York: Knopf, 1971.

anderthals under a "bed of flowers" became a key image in the new search for similarities between Neanderthal and human behavior and was picked up by Jean Auel in her prehistoric fiction series "Earth's children," first published in 1980.[22] Hochadel shows that these narratives, which could flourish due to the close interaction between archeological research and science pop culture, naturalized the comparison between prehistoric and modern life.

Rebecca Mertens analyzes the role of narrating for successful comparisons in the context of research management at the California Institute of Technology (Caltech) in the 1940s and 1950s. She claims that comparisons drawn between different research areas and their objects in the biological and chemical sciences gained their validation and persuasive power from what she calls "project organization narratives." Her case is the history of work pursued at Caltech's Chemistry and Biology Divisions, led by Linus Pauling and George Beadle, respectively. The basis of their joint work was the assumption that structural or spatial complementarity at a molecular level was the key to unlock the secrets of life. The successful model for this assumption was the antibody-antigen theory, which had gained traction already in the interwar period. Its success led the way to a systematic approach guiding many scientific and medical sub-specialties of the 1940s and 1950s into an era of collaborative research. Thus, the origin of the molecular life sciences can be seen much more directly in the paradigm of structural complementarity than in the arguably better known ideas of sequence complementarity and information that grew to dominance with the influx of physicists and the discovery of DNA's double helix structure. However, the role of structural complementarity kept being strong throughout this period, and was the base for the comparability of many different research agendas.

Support for the Caltech program in joint chemistry and biology of the molecular understanding of life came from the Rockefeller Foundation. Mertens's claim is that the Foundation did not only provide financial support but also the crucial narrative. Their program officers and science journalists were key actors in authoring the mentioned project organization narratives. The key part of the Rockefeller narrative in the immediate Post-WW II period was the story of lost opportunities during the war, the exploitation of basic research (which was deemed a limited resource), and the crucial role of making up lost ground by a concerted effort featuring collaborative work. Moreover, the essence of such successful research was to build the basis for

---

22   Auel, Jean M., *The Clan of the Cave Bear*, New York: Crown, 2011 [1980].

later application in industry, medicine, and society in general. Scientific comparative practice in a collaborative mode was thus made possible in the frame of a narrative of basic research, highlighting both its former shortcomings and its future promises.

Part III focuses on the material aspects of narrating and comparing with a special emphasis on the reception and historiography of art. The thought that epistemic practice is constituted not only by interconnected human activities, but also by instruments and objects that themselves develop a certain kind of material agency has been well developed in the course of the sociology of experimentation and laboratory practice.[23] For the recent history and sociology of culture techniques in the arts and the sciences, the relationship between material and human agency is a crucial subject of analysis.[24]

Joris Corin Heyder explores the relations between practices of seeing, comparing and narrating in early historical reconstructions of medieval art. The question of Heyder's analysis is how comparing images and aspects of images leads to the formation of what he calls a "narrative network," a "non-hierarchical interplay of actants and their narrative potential." With this approach, Heyder begins a new section in the present volume by introducing images into the interplay of comparisons and textual narratives. Starting from studying visual comparisons, Heyder includes paintings in the creation of narratives. He focuses on the role of medieval art in challenging the text-based view of the Middle Ages as a dark historical epoch. Heyder draws especially on the work of an eighteenth-century connoisseur and art critic Jean

---

23   See, for instance, Callon, Michel, and Bruno Latour, "Unscrewing the Big Leviathan: How Actors Macro-Structure Reality and How Sociologists Help them to do so." In *Advances in Social Theory and Methodology. Toward an Integration of Micro- and Macro Sociologies*, edited by K. Knorr-Cetina and A.V. Cicourel, Boston, London, Henley: Routledge & Kegan Paul 1981, 277-303. Latour, Bruno, *Science in Action. How to Follow Scientists and Engineers through Society*, Cambridge MA: Harvard University Press 1987. From a philosophical angle, Andrew Pickering has reflected on sociological concepts of human and material agency in: Pickering, Andrew, "The Mangle of Practice: Agency and Emergence in the Sociology of Science," *American Journal of Sociology* 99, 1993: 559-589. And likewise in his book: *The Mangle of Practice: Time, Agency, And Science*, Chicago: The University of Chicago Press, 1995. For the role of material agency in practice theory, see Schatzki, T., K. Knorr-Cetina, and E. Savigny, *The Practice Turn in Contemporary Theory*, London: Routledge, 2000.

24   See, for instance, Reckwitz, Andreas, "Toward a Theory of Social Practices: A Development in Culturalist Theorizing." *European Journal of Social Theory* 5(2), 200, 243–263. https://doi.org/10.1177/13684310222225432 [2019/02/12].

Baptiste Séroux d'Agincourt. This enables Heyder to scrutinize the impact of re-producing and re-presenting medieval art in the context of the Enlightenment, which has been notorious for depicting the Middle Ages as a mere link between antiquity and the Renaissance. Thus, while Séroux d'Agincourt's textual account follows the established narrative of the degenerate Middle Ages, the paintings chosen, assembled, and described create a different narrative that grants medieval art an independent role and impact that it did not have in previous historical works.

In challenging long established paradigms, such as the incompatibility of narrative and the pictorial, and in invoking practice theory and Actor Network Theory, Heyder establishes the concept of a visual-narrative network. He does so by analyzing in a step-wise fashion the practices of seeing, comparing, and narrating. Alluding to the analysis of the nineteenth-century art historian Franz Wickhoff, Heyder brings to the fore a visual narrative that relies on the basic mental practices of complementing, continuing, and distinguishing. Comparative arrangements of images, such as Séroux d'Agincourt's *Histoire de l'art par les monumens*, afford the opportunity to introduce the cultural persona of the beholder who creates stories even out of single images and forges connections between series of images. These are often based on comparing and lead to new comparisons, for example, by bringing in new *tertia comparationis* or standards of comparison. Here, the pictorial domain is more fruitful than the textual, there are "almost endless possibilities of identifying *tertia*." Of course, in the subsequent descriptions these *tertia* lead to new (textual) narratives: In Heyder's view, analysis of images and story-telling are creatively linked through practices of comparing. However, he ends with a note of caution. Each single configuration chosen by Séroux d'Agincourt can set free its own narrative, no sign-posts for generalization and abstraction are possible. But with this caveat attached, comparative image analysis can both enrich and limit narratives.

Britta Hochkirchen explores the relationship between comparing and narrating in the history of modern art. She examines how practices of comparing in art exhibitions supported the narrative of modern art. Hochkirchen specifies practices of comparing and narrating in written texts and curatorial activities in the historiography of modern art in the 1950s, the latter of which is exemplified by the first *documenta* in Kassel in 1955. Key to her argument is the analysis of comparisons in Werner Haftmann's canonical volumes *Malerei im 20. Jahrhundert* as well as in the exhibition space and their respective role in the temporal order of the narration of modern art in post-war

Germany. As Hochkirchen shows, both text-based and curatorial comparative practices created a narrative of modern art as a unified European project, establishing a temporal link between German art after World War I and the art in other European countries before and during World War I. However, the way in which comparisons were used for narration and also the narratives themselves differed decisively at the linguistic and the curatorial level of practice, even though Haftmann himself was responsible for the historical basis of the first *documenta* exhibition. Hochkirchen explains the different narrative strategies by distinct rationales of temporal ordering and experiences: In Haftmann's written text, the narrative of modern European art was created by means of its *distinction* from the old mimetic style of mirroring reality, attributed to the period of the Renaissance, on the one hand, and from the contemporary art style of the Soviet Union, on the other. Thus, discontinuity played a crucial role in Haftmann's narrative of the development of modern art in Europe. However, the exhibition entirely focused on the *continuity* of the abstract mode of art works within the European context of modern art. The arrangement of the exhibition pieces (e.g., Picasso's *Girl before a Mirror* and Winter's *Komposition vor Blau und Gelb*) encouraged a comparative view emphasizing their similarity with respect to the "progress" of the abstract mode.

In this third part we go beyond the sciences and explore the role of narratives and comparisons in the humanities. Interestingly enough, both contributions discuss how comparisons may produce narratives. Thus, they invert the transition from narrating to comparing that has been expounded as the dominant mode in the sciences. In reconstructing the historical evolution of art, comparison is the means to producing narratives. Such narratives play the role of explanations in that they make sense of the similarities and differences exhibited. Whether this feature happens to come up only in these two contributions to this volume and is thus peculiar to these cases or whether it is generalizable to a wider realm remains an open question.

The volume has grown out of the workshop "Practices of Comparing and Narrating in the Sciences" within the collaborative research center on Practices of Comparison (SFB 1288, Praktiken des Vergleichens). The workshop took place at Bielefeld University in April 2018 and was organized by Veronika Hofer. We thank Veronika for her continued effort.

Does Narrative Matter?
Engendering Belief in Electromagnetic Theory

M. Norton Wise

> What is the use then of imagining an electro-tonic state of which we have no distinctly physical conception, instead of a formula of attraction which we can readily understand? I would answer, that it is a good thing to have two ways of looking at a subject, and to admit that there are two ways of looking at it.
>
> J. C. Maxwell, "On Faraday's Lines of Force" (1855)[1]

With these words James Clerk Maxwell positioned himself with respect to the sharply differing perspectives on electromagnetic action that were occupying natural philosophers by the time he published his first paper on the subject in 1855. How should they think about the action between two wires carrying electric currents. Should they imagine an action mediated by a magnetic field in all space describable in terms of "lines of force" and an electro-tonic state existing at every point: "of which we have no distinctly physical conception." Or should they suppose the space itself to be empty and imagine instead a direct unmediated action between moving electric particles (atoms) constituting currents: captured by a mathematical formula "which we can readily understand." This famous conundrum raises the question of how physicists at the time could compare the "field theory" of Maxwell with the "action at a distance theory" of Wilhelm Weber.[2]

---

1 Maxwell, James Clerk, "On Faraday's Lines of Force." In *Transactions of the Cambridge Philosophical Society 10*, Part I, 1855, in *The Scientific Papers of James Clerk Maxwell*, Cambridge: Cambridge University Press, 1890, 155-229, on p. 208.
2 For succinct and insightful but more technical discussions of Weber, Maxwell, Faraday, and others appearing below see Darrigol, Olivier, *Electrodynamics from Ampère to Einstein*, Oxford: Oxford University Press, 2000.

One way to look at the problem of comparison is in terms of believability. How did people come to believe in one conception or the other? Apparently the usual criteria of empirical validity, mathematical coherence, and comprehensiveness were not enough, since in this case both representations seemed capable of encompassing all relevant phenomena. It was more nearly a matter of belief in one sort of imagined "world" versus another. And in this situation how the imagined world was narrated was important. In order to develop this perspective I will consider an analogy with the function of narrative in supporting belief in Greek mythology, largely following a recent analysis by Sarah Iles Johnston.[3]

Note: Narratologists often think of narrative as defined by an unfolding in time of a connected sequence of events. I use it here in the broader sense of an unfolding of a representation or interpretation of a part of the world, without any necessary reference to temporality. See my concluding comment below.

1.   Two Conceptions of Electromagnetic Action

Before entering directly on the topic of how narratives support belief I will first describe in Part I, more or less for themselves, Maxwell's presentation of field theory in terms of Faraday's "lines of force" and the electro-tonic state and Weber's presentation of action at a distance between particles, while pointing to some of their narrative characteristics. I will then in Part II broaden the discussion to include more general considerations of how narrative supported belief within a "story world," using Johnston's categories as adapted for the examples of Michael Faraday's *Experimental Researches* for field theory and Gustav Theodor Fechner's *Atomenlehre* for action-at-a-distance.

---

3   Johnston, Sarah Iles, *The Story of Myth*, Cambridge MA: Harvard University Press, 2018. Johnston, Sarah Iles, "Narrating Myths: Story and Belief in Ancient Greece," *Arethusa* 48 (2), 2015: 173-218. Johnston, Sarah Iles, "The Greek Mythic Story World," *Arethusa* 48 (3), 2015: 283-311.

1.1. Maxwell, "On Faraday's Lines of Force" (1855)

In the first thirty three pages of a seventy six page paper Maxwell carefully unfolded verbally a picture of how lines of electric and magnetic force could be represented in familiar terms as lines of fluid flow, as depicted in figure 1.

Figure 1: Lines of force surrounding a bar magnet with north and south poles.

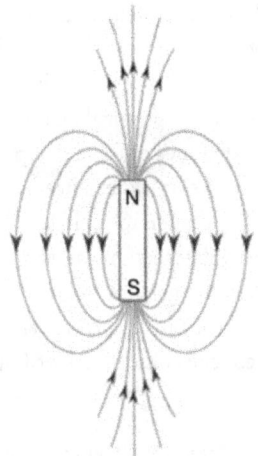

The first fifteen pages of this discursive narrative contained no mathematics at all while the next eighteen employed just the simplest algebra. It was only with an intuitive image established that he would then develop in twenty pages a set of formal equations that might govern the interaction of electric and magnetic lines in terms of Faraday's electro-tonic state. A summary of the entire structure in six laws completed the account, with examples of their application.

This is the earliest instance of Maxwell's famous use of "physical analogies": "my aim has been to present the mathematical ideas to the mind in an embodied form, as systems of lines and surfaces and not as mere symbols, which neither convey the same ideas, nor readily adapt themselves to the phenomena to be explained."[4] It would be a mistake to think here of "embodied mathematics" as a purely intellectual affair, in which mathematical expressions simply receive concrete exemplification in a physical process. It is certainly that but much more. Repeatedly through his life Maxwell emphasized that embodiment was also a matter of awakening the senses. As he would put it in his Presidential Address to the British Association in 1870, "[many physicists] calculate the forces with which the heavenly bodies pull at one another and they feel their own muscles straining with the effort. To such men momentum, energy, mass are not mere abstract expressions of the results of scientific inquiry. They are words of power, which stir their souls like the memories of childhood."[5] It is helpful to keep this highly sensual aspect in mind when thinking of how Maxwell sought to embody the lines of force and their dynamical behavior in a narrative. He wanted to bring them to life like "the memories of childhood," or perhaps the characters in a short story. In the embodied mathematics of a physical analogy he aimed to conceptually integrate diverse aspects of the lines of force perspective while preserving the "vividness" and "fertility" of sensory experience.[6]

To that end he asked his reader to "consider these curves not as mere lines, but as fine tubes of variable section carrying an incompressible fluid."[7] Beginning from the simplest images, immediately accessible to anyone who had seen water flowing, whether in a stream or simply in a basin, Maxwell unfolded the geometrical conception of lines of flow in a three-dimensional

---

4   Maxwell, "Faraday's Lines," 156, 187. The literature is immense. For the specific religious and cultural context in which Maxwell developed his use of physical analogy see Lambert, Kevin, "The Uses of Analogy: James Clerk Maxwell's 'On Faraday's Lines of Force' and Early Victorian Analogical Argument," *British Journal for the History of Science* 44, 2011: 61—88. On the method of reasoning see Cat, Jordi, "On Understanding: Maxwell on the Methods of Illustration and Scientific Metaphor," *Studies in History and Philosophy of Science, Part B: Studies in History and Philosophy of Modern Physics* 32, 2001: 395-441. Nersessian, Nancy, "Maxwell and 'the Method of Physical Analogy': Model-based Reasoning, Generic Abstraction, and Conceptual Change." In *Reading Natural Philosophy: Essays in the History and Philosophy of Science and Mathematics*, edited by D. B. Malament, Chicago: Open Court, 2002, 129-166. Generally see Darrigol, *Electrodynamics from Ampère to Einstein*, 137-147.
5   Maxwell, James Clerk, "Address to the Mathematical and Physical Sections of the British Association," *Report of the British Association for the Advancement of Science* 40, 1870, 215-229, on p. 220.
6   Maxwell, "Faraday's Lines," 156.
7   Ibid., 158.

space, moving from lines to tubes of flow and gradually adding conditions on velocity, sources and sinks, a resisting medium, pressure gradients, and changes of medium. The result was an accessible image of a space full of flowing fluid, which, although not initially developed mathematically, was easily expressible in mathematical terms.

To put it a bit differently, lacking any physical theory of what a field of force might be, Maxwell led his reader into a fictional world containing a "purely imaginary substance," which exhibited the properties he sought. "It is not even a hypothetical fluid which is introduced to explain actual phenomena. It is merely a collection of imaginary properties which may be employed for establishing certain theorems in pure mathematics in a way more intelligible to many minds and more applicable to physical problems than that in which algebraic symbols alone are used."[8] Through this conceptually enriching if fictional narrative, rendered in everyday terms, he sought to stimulate the reader's imagination, giving almost sensory existence to the idea of lines of force as analogous to lines of fluid flow.

Having established his basic image in these familiar terms Maxwell easily employed it to draw together nearly all of the phenomena of electricity and magnetism as conceived by Faraday, replacing the idea of attraction at a distance with lines of force conducted through space, including: the distribution of electric lines around positive and negative charges of static electricity; the distribution of magnetic lines around north and south poles of magnets (figure 1); the distribution of electric current lines in a conductor; and the equivalence of electric currents and magnets in electromagnetism (so that a small electric circuit behaved exactly like a small bar magnet). The existence of electromagnetism meant that the two systems of electric current lines and magnetic lines, each conceived separately in terms of flow, had to be interrelated dynamically, so that the properties of each system could be understood in terms of the properties of the other. Their qualitative relation can be readily understood with reference to a coil of wire carrying a current, which behaves like a bar magnet with north and south poles and produces an equivalent distribution of magnetic lines (figure 2a).

---

8   Ibid., 160. As Lambert, "Uses of Analogy," 86, puts it, "Maxwell thought the manipulation of objects could also discover ideas."

Figure 2: (a) A current-carrying coil behaves like a bar magnet. (b) An electric current line and a magnetic line are related like a "mutual embrace."

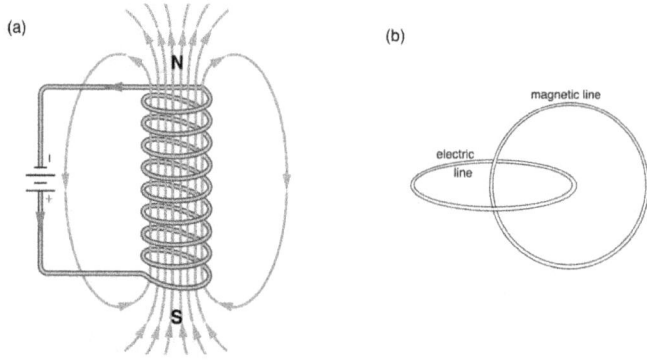

The pattern of the magnetic distribution by itself can be seen as a dynamic balance, resulting from a tendency of each line to *contract* along its length and for adjacent lines to *repel* each other laterally. But these effects are mirrored reciprocally by the tendency of the current lines (or turns in the coil) to *extend* along their length and for adjacent lines to *attract* laterally.

With his flair for evoking sensory perception Maxwell labelled Faraday's image of these interlocked rings the "mutual embrace" of electricity and magnetism (figure 2b).[9] He had at hand no physical analogy that could account for the interrelation of the lines but his flow analogy did provide key concepts of flow velocity and pressure gradient at any point, or "quantity" and "intensity" of the flow, in terms of which the reciprocal dynamics might be represented mathematically. The picture of mutual embrace suggested that just as the *quantity* of current passing through a surface surrounded by a magnetic line could be expressed in terms of the *intensity* in the magnetic line, so the *quantity* of magnetic force passing through a surface enclosed by a current line should be expressible in terms of the current's *intensity*. But no

---

9   Maxwell, "Faraday's Lines," 184, 194f. From Faraday, *Experimental Researches*, III, 3265 and plate IV, fig. 1. For Maxwell's continuing use of the metaphor in later papers see Wise, M. Norton, "The Mutual Embrace of Electricity and Magnetism," *Science* 203 (4387), 1979: 1310-1318.

such relation of magnetic quantity to current intensity existed. Thus mutual embrace remained a highly suggestive image, to which Maxwell had led his reader through an illuminating flow analogy for lines of force, but it ended up showing that the story he had constructed was as yet incomplete.

This inadequacy was particularly troubling for Faraday's great discovery of electromagnetic induction, whereby an increase or decrease of the magnetic quantity passing through a surface surrounded by a closed conductor would induce a current in the conductor. Like Faraday, Maxwell thought there must be some corresponding condition in the conductor, an "electro-tonic state," which was responsible for the current. But lacking any physical analogy with which to embody this speculation, it remained a puzzling element within the picture of lines of force. He therefore turned in the second half of his paper to a purely mathematical representation of the electro-tonic state. In this abstract form it served nearly to complete mathematically the symmetry of the mutual embrace while also encompassing electromagnetic induction. But it remained a somewhat ghostly stranger in Maxwell's integrative narrative. He left his reader with the hope that an extended physical analogy would someday complete the picture. "By a careful study of the laws of elastic solids and of the motions of viscous fluids, I hope to discover a method of forming a mechanical conception of this electro-tonic state adapted to general reasoning."[10] This aim to develop a more complete narrative, which did not depend in the first instance on mathematical expression, would guide Maxwell's development of electromagnetic field theory for many years.

1.2   Weber, Elektrodynamische Maassbestimmungen (1846)

In sharp contrast to Maxwell's aim of physical embodiment of mathematical relations, Wilhelm Weber sought an abstract mathematical relation that would provide a *Grundgesetz* for all electrical action, where the term *Grundgesetz* implies a foundational law governing the constitution of the phenomena and from which they can be derived. And while Maxwell approached his subject as a reflective theorist looking for a new conceptual structure, Weber presented himself as a rigorous experimentalist seeking quantitative empirical grounding for a generalized law of action at a distance, a law that would do nothing more than express the results of his measurements, thus the title

---

10   Maxwell, "Faraday's Lines," 188.

"Electrodynamic Measurements" (or Determinations of Electrodynamic Measure).[11]

As such Weber's 170 page essay has a structure very different from Maxwell's. He let his reader know from the beginning that there was a character behind the scenes that would ultimately appear as a central figure, namely electric currents represented in terms of positive and negative particles of electric fluids flowing in opposite directions inside a conductor. But these particles did not immediately concern him. Instead he began his narrative from the closest expression yet attained to what he called a "fundamental law" of the force acting between two current-carrying wires (not the flowing electric fluids themselves). The French mathematical and experimental physicist André-Marie Ampère had succeeded in expressing this law as an action at a distance between any two infinitesimal elements of the wires, depending on their current strengths, relative orientations, and the inverse square of the distance between them.[12] But to Weber, Ampère's accomplishment had a great weakness. He had not actually been able to measure the force acting between two current-carrying wires. Instead he had relied on so-called null experiments, reasoning for particular arrangements of currents that if no effect were observed then the force had to have the form he ascribed to it. Although justly famous, Ampère's method could neither give positive measurements of the forces nor establish absolute values of the currents. He simply did not have the necessary instruments.

Weber had the solution. He devoted the first hundred pages of his book to the design and operation of a new "electrodynamometer" of extraordinary precision.[13] From a literary perspective Weber's presentation of his instrument was itself a work of considerable rhetorical skill, another narrative unfolding of a vivid image, but this time of the creative design, operation, and uses of the key component—the key actor—in an empirically based narrative that would ultimately lift Ampère's "fundamental" law of action at a distance between current-carrying wires into a proper *Grundgesetz*. Drawings

11   Weber, Wilhelm, *Elektrodynamische Maassbestimmungen*, Leipzig: Weidmann'sche Buchhandlung, 1846.

12   Ampère, André-Marie, "Mèmoire sur la théorie mathématique des phénoménes electrodynamiques uniquement déduite de l'expérience," *Mémoires de l'académie royale des sciences de l'institut de France 6*, 1823 : 175-388. On Ampère's theoretical and experimental methods see Hofmann, James R., "Ampère, Electrodynamics and Experimental Evidence," *Osiris 3*, 1987: 45-76. Darrigol, *Electrodynamics from Ampère to Einstein*, 6-13, 23-30.

13   Darrigol, *Electrodynamics from Ampère to Einstein*, 54-66.

were critical to the reader's appreciation of the arrangement of components and of how they functioned (figure 3).

In its basic version, an outer fixed coil of current-carrying wire sur-rounded an inner moveable coil, which was placed perpendicularly to it and was suspended on a pair of fine wires for sensitive detection of any rotation produced by action between the coils. A small mirror mounted on the inner coil allowed tiny movements to be read by reflection through a telescope on a scale placed six meters away.[14] The reader's initial appreciation for the refinement of the instrument and its capacities, however, was built not only on detailed description but on Weber's story of its origins, specific identification of the instrument maker who perfected it, extensive calibration data, analysis of precision, and sources of error. Fully fleshed out in this way, the electrodynamometer functioned as the trusted agent of truth in Weber's account.

Only having established this material foundation did Weber return to his reworking of Ampère, measuring with precision and with named witnesses to the observations the action between the current-carrying coils of his elec-trodynamometer. The result completely confirmed Ampère's fundamental law of the force acting at a distance between current elements. He then turned to Faraday's discoveries of current induction to show that the electrodynamome-ter could similarly confirm those results, both qualitatively and quantitatively. At this point in his narrative it would seem that Weber had not only presented his instrument as an agent capable of reworking experimentally all known phenomena of electrodynamics but had made the electrodynamometer into an instrument that in effect reified those phenomena as results of action at a distance.

Nevertheless a major difference existed between the Ampère and Faraday results, for while Ampère's law referred to electric currents, the force it actu-ally described acted on the conductors carrying the currents. In this sense, it was not an electrical force at all. Faraday's induction of currents, on the other hand, concerned a force acting on the electricity itself inside a conductor to create a current. That distinction opened the door to the second half of We-ber's essay, in which he revived the background image of electric fluids that he had originally only mentioned. He now sought a general law of truly electrical

---

14   Weber adapted the bifilar suspension and telescopic mirror reading technique from
      Gauss's magnetic measurements, on which he collaborated. Weber, *Elektrodynamische
      Maassbestimmungen*, 10.

Figure 3: Wilhelm Weber's Electrodynamometer. Weber, Elektrodynamische Maassbestimmungen, 11.

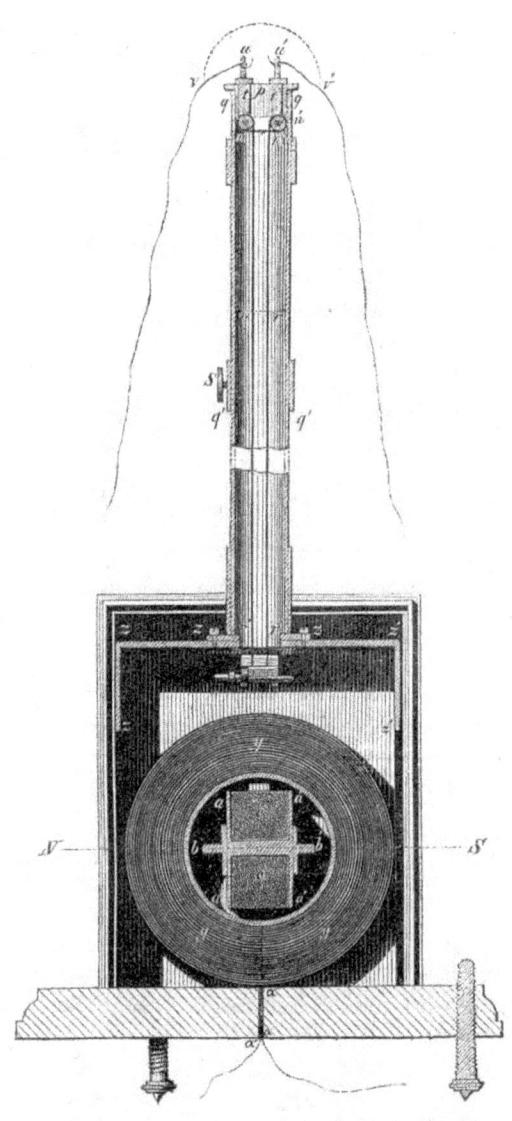

forces acting between masses of positive and negative electricity (effectively electric point atoms, as for Fechner below). Returning to the assumption that currents consisted of positive and negative fluids moving inside conductors, he asked what supplement of the familiar inverse square law ee'/r^2, which governed the electrostatic force between electric masses e and e' at rest with a distance r between them, would apply if the masses were in relative motion, as in a wire carrying a current (figure 4).

Figure 4: Weber's law of force between electric particles e and e' flowing in a wire carrying current: $F = ee'/r^2 (1 - k^2v^2 + 2kra)$

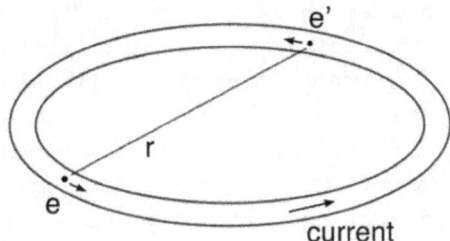

current

From looking at only two facts about the Ampèrian forces between current elements he quickly inferred that the simplest supplement of the electrostatic law would be two additional terms, one depending on the square of the relative velocity v between the electric masses and a second depending on their relative acceleration a:

$$F = (ee'/r^2)(1 - k^2v^2 + 2kra),$$

where k is a constant. With equal facility Weber showed from a single fact about Faraday's induction of currents that it also fit this abstract law, confirming its validity.

It may not be immediately obvious just how dramatic this result was. Nothing in the preceding 100 pages of presentation and legitimation of the electrodynamometer had prepared the reader for a simple mathematical expression that subsumed all of electrostatics and electrodynamics in one law of force for electric masses. A few pages of skillful reasoning had converted a tour de force of experimental prowess into a formula that provided the calculational basis of all electrical action. After one more generalizing move (a mathematical transformation of Ampère's law into the new law for electric masses), Weber reached the climax of his narrative. He could now call his accomplishment the "*elektrische Grundgesetz*", the law of constitution of any and all electrical phenomena.[15] It remained only to prove that in fact the electrical phenomena could be formally derived from the *Grundgesetz*, including of course all of the refined measurements made by the electrodynamometer for both Ampère's constant currents and Faraday's induced currents.

But Weber's *Grundgesetz* was a law like no other. That the force between two bodies should depend on their relative velocity and acceleration, or should be time-dependent, challenged basic assumptions of mechanics.[16] Nevertheless Weber pressed on, suggesting that other forces too, such as gravitation, might have to be similarly supplemented. "*A priori* this question cannot be decided, because formally in the assumption of such forces there is neither any contradiction nor anything unclear or indeterminate." Furthermore, the purpose of such "fundamental laws" was not "to give an *explanation* of the forces from their true grounds but only to give ... a useful general method for *quantitative* determination of the forces according to the fundamental measures determined in physics for space and time."[17] The *Grundgesetz* suggested even that multibody forces might exist, since the acceleration between two masses could be affected by a third, as in recently discovered catalytic forces of chemistry. Indeed, mediating effects of an ether might be contemplated, as Faraday's recent discovery of magnetic rotation of the plane of polarization of light suggested.[18] Thus a whole new world of possibilities opened up.

---

15   Weber, Elektrodynamische Maassbestimmungen, 119.

16   For the immediate controversy see Bevilacqua, Fabio, "Theoretical and Mathematical Interpretations of Energy Conservation: The Helmholtz-Clausius Debate on Central Forces 1852-54." In Universalgenie Helmholtz: Rückblick nach 100 Jahren, edited by L. Krüger, Berlin: Akademie Verlag, 1994, 89-106. Darrigol, Olivier, "Helmholtz's Electrodynamics and the Comprehensibility of Nature." In Universalgenie Helmholtz, 216-242.

17   Weber, Elektrodynamische Maassbestimmungen, 112-113.

18   Ibid., 168-170.

But Weber wanted to be clear that the compelling picture he presented of direct action at a distance between electric masses, was a fictional, if realistic, construction. Concerning currents: "The simultaneous movement in opposite directions of positive and negative electricity ... may in reality not exist at all, but for our purpose may be regarded as an *ideal* motion, which ... [for] actions *at a distance*, may *represent* the motions really present."[19]

In summary, and somewhat like Maxwell, Weber built up an experimental and theoretical narrative that would launch a generalized concept of action at a distance, in which forces could be time dependent. The conception was highly successful at drawing together disparate elements, even if fictional. The basic object of understanding on this view was a pair of particles, or electric atoms, between which a force acted. The force itself was an abstract relation in space and also time: "because a time-dependent relation is just as measurable a quantity as distance."[20] In contrast to Maxwell, however, the space surrounding the two atoms contained nothing: no force, no field, and of course no lines of force.

2. Believability and the Techniques of Narrative

Both Maxwell and Weber carefully structured their narratives of electromagnetic phenomena to make the unfamiliar familiar and to yield a climactic moment in which a strange new object emerged. For Maxwell the story culminated in an electro-tonic state, which had never been observed and for which he could provide no ordinary physical conception but only a suggestive mathematical symmetry. For Weber the culmination was a time-dependent force, whose violation of established principles Weber countered with appeals to logical validity and to possible extension to other areas, such as catalytic action.

Thinking of these fictional constructions in rhetorical terms, my question now is what made them believable in everyday terms. This is the same question that classicist and historian Sarah Johnston has asked for Greek mythology: "how, exactly, does the narration of myth sustain a metaphorical connection between the mythic and quotidian worlds."[21] One aspect jumps out

---

19   Ibid., 100.
20   Ibid., 113.
21   Johnston, *Story of Myth*, 79.

immediately. Both Maxwell and Weber spent the majority of their presentations making the reader feel at home within the worlds they were in the process of building, well before they revealed their creative fictions. Maxwell did this through the flow analogy, which was accessible to anyone who had paid close attention to fluid flow. Weber did it through his extended presentation of the design, operation, and measurements of the electrodynamometer, all of which confirmed Ampère's and Faraday's laws in terms of action at a distance. Only after having gone to considerable length to establish this familiarity and normalcy—and thereby their own legitimacy and a suspension of disbelief—did they guide their readers into consideration of a possible expanded reality.

Techniques of this kind for introducing the fictional or extraordinary into the quotidian are so common in narratives dealing with otherwise questionable events or beings that it has been designated the "X/Y Format"—X for the familiar and Y for strangeness—by the sociologist Robin Wooffitt.[22] It is only one of many techniques, however, that Johnston has highlighted in skillfully constructed narratives, which contribute to the believability of the gods and heroes of Greek myths.[23] It is not that speculative stories about electromagnetism are much like myths—lines of force and electric atoms are characters of a different sort from Heracles or Theseus—but the techniques of narration that enhance their believability are similar. Among those techniques (but adapted and reordered) I will take up the role of: *conceptual metaphor, serial narration, multipliers* (Johnston's *plurimediality*), and *story world*. Together they help to clarify the *pragmatic effect* of effective narration. To explore this view for audiences of electromagnetism I will move out from the highly focused representations by Maxwell and Weber to the broader narratives of Faraday and of Gustav Fechner.

A key aspect of Johnston's entire discussion of the effectiveness of techniques of narration is her treatment of emotional and cognitive responses as integrally related. Although I will not explicitly take up that relation here, Maxwell's view of the sensory role of physical embodiment of mathematical

---

22   Johnston, *Story of Myth*, 98-102; Wooffitt, Robin, *Telling Tales of the Unexpected: The Organization of Factual Discourse*, Hemel Hempstead: Harvester Wheatsheaf, 1992, 114-152.

23   Mayer, Adrienne, *Gods and Robots: Myths, Machines, and Ancient Dreams of Technology*, Princeton: Princeton University Press, 2019 is also highly relevant here for its accounts of the relation of fictional automata in Greek myths to familiar technology, with believability, and also creativity, running in both directions.

formulas can serve as a reminder of its importance, which reappears below for Fechner.[24]

2.1 *Faraday*, Experimental Researches in Electricity *(1831-1852)*

Over the course of twenty years from 1832 to 1852 Michael Faraday published in the *Philosophical Transactions of the Royal Society* and other journals the articles that would make up the three volumes of his *Experimental Researches in Electricity*. Having made his reputation with major discoveries in chemical equivalents and electrochemistry he had turned to electricity and magnetism proper. The *Researches* contained an astonishing collection of discoveries, including electromagnetic induction (1831), specific inductive capacity (1837), diamagnetism (1845), magnetic rotation of light (1845), and many others of both theoretical and practical significance. Throughout these works Faraday continued to ponder and to develop the idea of lines of force as an alternative to action at a distance.[25]

Conceptual metaphor. The term "lines of force" functioned during this development as what Johnston, borrowing from the linguists George Lakoff and Mark Johnson, calls a *conceptual metaphor*. Such metaphors, she observes, commonly functioned in the narration of Greek myths to figuratively connect events in the everyday world to events in the world of the myth and thereby support belief.[26] In Faraday's case, his use of lines of force as a central metaphor not only connected many different strands in the actual world of his laboratory experiments (as in figure 5), but connected them as well to an imagined world in which forces had something like material status.

In retrospect, Faraday's metaphorical language might seem to have been highly effective. It is well to remember, however, that it was not necessarily so, especially among those who prioritized mathematical expression. William Thomson, for example, who would ultimately become Faraday's first great mathematical interpreter, when originally encountering Faraday's language

---

24   Johnston, *Story of Myth*, e.g. 10, 66-67, and throughout.
25   Faraday, Michael, *Experimental Researches in Electricity*, 3 vols., facsimile reprint, London: Quaritch, 1855, cited by paragraph number. On the sources and significance of Faraday's use of lines of force see Gooding, David, "'Magnetic Curves' and the Magnetic Field: Experimentation and Representation in the History of a Theory." In *The Uses of Experiment*, edited by D. Gooding, T. Pinch, and S. Schaffer, Cambridge: Cambridge University Press, 1989, 183-223. Darrigol, *Electrodynamics from Ampère to Einstein*, 16-22, 31-41.
26   Johnston, *Story of Myth*, 67, 73.

Figure 5: Faraday's image of iron filings mapping the lines of force around two circular magnets with north and south poles. Faraday, Experimental Researches, III, Plate IV, Fig. 4.

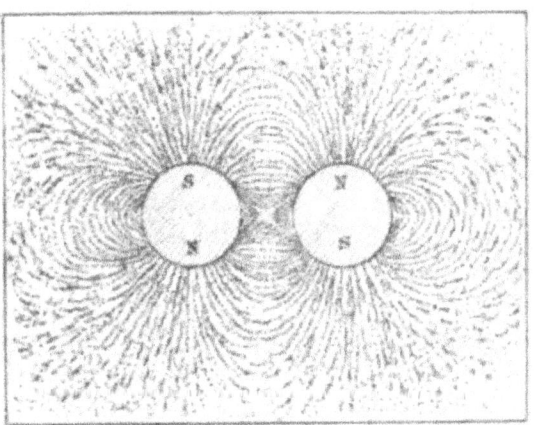

of electrostatic "induction in curved lines" in 1843, wrote that "I have been much disgusted with his way of *speaking* of the phenomena, for his theory can be called nothing else."[27] It would be two years before he fully appreciated that Faraday's "way of speaking" fit quite well with his own development of a mathematical analogy between heat conduction and electrostatic action, with which he had shown their near mathematical equivalence. Thomson's analogy between flux of heat and lines of force would provide Maxwell's starting point for his own fluid flow analogy ten years later. The seemingly so obvious power that we see today in Faraday's conceptual metaphor is actually a product of historical recountings, not unlike the way in which repeated narration and performance of Greek myths around conceptual metaphors enhanced the believability of Gods and heroes.

Serial narration. Closely related to this historical aspect of effective metaphors, but within Faraday's own reports of his experiments, is their *serial narration.* The articles in the three volumes were narrated serially over

27   Smith, Crosbie, and M. Norton Wise, *Energy and Empire: A Biographical Study of Lord Kelvin*, Cambridge: Cambridge University Press, 1989, 213.

twenty years. The seriality was quite literal, with episodes appearing at irregular intervals with a series number and in numbered paragraphs.[28] Johnston takes serial narration to have been another of the important factors contributing to belief in myths. Offered up in small installments, each with its own focus but always contributing to a single story line, the series encouraged readers to contemplate each episode in relation to previous ones and in anticipation of what might appear next, as though following one of Charles Dicken's serialized novels or a TV series like *Downton Abbey*. Faraday encouraged such responses with many back-references and suggestive speculations about future developments, as in the following excerpt from the Eleventh Series (1837).[29]

> 1163. In the long-continued course of experimental inquiry in which I have been engaged, this general result has pressed upon me constantly, namely, the necessity of admitting two forces, or two forms or directions of a force (516. 517.).
>
> 1164. When I discovered the general fact that electrolytes refused to yield their elements to a current when in the solid state, though they gave them forth freely if in the liquid condition (380. 394. 402.), I thought I saw an opening to the elucidation of inductive action, and the possible subjugation of many dissimilar phenomena to one law.

As this excerpt also suggests, seriality offers another interesting mode of reading, namely, reading out of sequence, so that readers are able continually to reconstruct the back-story for themselves. Such reconstruction can suggest different approaches and new insights, enhancing personal engagement. Johnston argues that all of these aspects of serial narration give characters a life of their own, which in itself contributes to their believability.[30]

Multipliers. Similar effects follow from various means of multiplication, whether by different authors, different outlets, different voices, or different

---

28   For extended discussion of "Seriality and Scientific Objects in the 19th Century" see the special double issue of *History of Science*, 48 (2010), edited by Nick Hopwood, Simon Schaffer, and James Secord.

29   Faraday, *Experimental Researches*, I, 1163-1164. As Faraday scholar Geoffrey Cantor has observed (verbal comment) it may be significant for reader response that Faraday's installments appeared about a year apart rather than only a week apart, as for Dickens.

30   Johnston, *Story of Myth*, 32, 91-96, 246-252. I am here collapsing the distinction of "series" from "serial" in episodic narration.

media. Johnston develops this as *plurimediality*.[31] Although most of Faraday's articles appeared in the prestigious *Philosophical Transactions*, for example, he placed some of them in the more widely read *Philosophical Magazine* and in the popular *Proceedings of the Royal Institution*, while preserving the numbered ordering of the serial narration. These different outlets not only multiplied his audience; they also presented his work with different degrees of speculative freedom and different levels of technicality.

Looking more widely, a considerable variety of authors contributed to the diversity of specific meanings and contexts that informed Faraday's lines of force. The chemist John Frederic Daniell dedicated his *Introduction to the Study of Chemical Philosophy* to giving an elementary view of Faraday's philosophy, including the mediating action of lines of force. There Thomson encountered the claims for "curved lines," which he initially considered nothing but verbiage but soon elaborated mathematically through his analogy to heat conduction.[32] And while Thomson admired Maxwell's similar use of physical analogy, he always rejected Maxwell's introduction of Faraday's electronic state from mathematical symmetry alone. Thus Daniell, Thomson, and Maxwell (among others) served as multiple narrators of the lines of force, whose differing interpretations contributed to the sense of their underlying reality. Other multipliers included the use of different modes of expression for the purpose of skillful narration, most prominent here being the mix of verbal, mathematical, and visual means that different authors used to capture Faraday's experiments and his already highly visual language.

Story world. Conceptual metaphors, serial narration, and multipliers of various kinds work together to create what Johnston and others call a *story world*. On entering the story world of Greek myths, we become familiar with a collection of characters whose stories become intertwined with each other. It is not so important that they appear always with the same personalities but that they create a dense network of relationships.[33] And so it was in Faraday's world of lines of force. Always exploring the possibilities for a reality in which forces are more substantial and fundamental than matter itself, he regularly repeated the view that forces of all kinds are expressions of one force and

---

31   Johnston, *Story of Myth*, 27-28, 156-176. Plurimediality takes the identity of the object of narration outside any particular author or presentation.

32   Daniell, John Frederic, *Introduction to the Study of Chemical Philosophy*, 2nd ed., London: Parker, 1843, 255-256.

33   Johnston, *Story of Myth*, 25-26, 121-146, as network 131-139. See also, Johnston, "The Greek Mythic Story World," 283-311.

are convertible one into the other.[34] His overarching narrative thus aimed at the ultimate goal of interrelating chemical reactions and heat with electricity, magnetism, light, and even gravity. Concerning the interlocked rings of electric and magnetic lines of force (figure 2b), which Maxwell would call their "mutual embrace," he offered: "their relation ... probably points to the intimate physical relation, and it may be, to the oneness of condition of that which is apparently two powers or forms of power, electric and magnetic."[35] Similarly, with respect to the magnetic rotation of light, he remarked: "Thus is established ... a true, direct relation and dependence between light and the magnetic and electric forces; and thus a great addition made to the facts and considerations which tend to prove that all natural forces are tied together, and have one common origin (2146.)."[36] Within this developing story world each of the topics and each of the installments of Faraday's long series of *Experimental Researches* became intertwined with the others through lines of force and each gained credibility from its place in the network in relation to the others.

Pragmatic effect. All of the techniques of effective narration that I have briefly described contributed to the believability of Faraday's conception of how forces functioned in the world. When successful, these techniques made the elusive notion of lines of force seem as real as wires and inspired his followers to try out the experiments for themselves, enlivening the ideas with their own experience, which Faraday always encouraged. Others formulated their own work in corresponding terms. Thomson and Maxwell are the obvious examples. This capacity of narration to affect how others think and act has been called the pragmatic effect.[37] Although the term might be applied to many forms of presentation, it refers here specifically to the capacity of an audience to introduce entities from a story world into their real world without an overly strained sense of fiction, having acquired a new openness to possible realities. Perhaps the most difficult of those realities in Faraday's narrative of lines of force was the electro-tonic state. As Faraday himself put it: "Again and again the idea of an electro-tonic state (60. 1114. 1661. 1729. 1733) has been forced on my mind; such a state would coincide and become identified with that which would then constitute the physical lines of magnetic force."[38] On

34   Faraday, *Experimental Researches*, III, e.g., 57, 366, 376, 877, 961, 2071, 2146.
35   Ibid., 3268.
36   Ibid., 2221.
37   Johnston, 20- 21, 57-58, 76-80, citing work of Claude Calame.
38   Faraday, *Experimental Researches*, III, 3269.

entering into Faraday's story world, Maxwell—but not Thomson—acquired a similar sense of the almost necessary reality of the imagined state. I have suggested that this was in part the pragmatic effect of effective narration. Maxwell responded by enriching the story world with his own physical analogy for lines of force and then reintroducing the electro-tonic state mathematically, as yet without any physical conception of it but with the expectation that it would soon appear in a prominent role.

2.2   Fechner, Atomenlehre (1855)

In order to obtain a similarly broad view of the believability of Wilhelm Weber's *Grundgesetz* in narrative terms it will be instructive to consider the work of Gustav Theodor Fechner.[39] The Leipzig physicist and philosopher was already a prominent intellectual who had published essays, books, and poetry, on everything from life after death to the mental life of plants, when in 1855 his sweeping tract on the atomistic conception of the world appeared, written in a distinctly literary vein and using Wilhelm Weber's work as the lynchpin of the presentation. Fechner had suffered a debilitating mental collapse in 1839, which effectively blinded him and which led to Weber assuming his professorship at Leipzig from 1843 to 1849, where they interacted closely.[40] Fechner had been pursuing an atomistic view of nature since the 1820s and in 1845 he published a partial account of the relation of Faraday's induction to Ampère's law of currents, modeling a current as equal and opposite motions of positive and negative electric masses. There he was able to announce that Weber had actually succeeded in subsuming all electrical phenomena under a single law of force.[41]

But Fechner had a much more ambitious agenda, one in which physics melded into philosophy and psychology and all three into "psychophysics," for which he is best known. It was the relation of physical and mental states

---

39   For a comprehensive analysis of Fechner's work, which informs my discussion here, see Heidelberger, Michael, *Nature from Within: Gustav Theodor Fechner and his Psychophysical Worldview*, trans. C. Klohr, Pittsburgh: University of Pittsburg Press, 2004.

40   Weber had himself been dismissed from his professorship at Göttingen in 1837 as one of the political protesters known as the "Göttingen Sieben."

41   Fechner, Gustav Theodor, *Maassbestimmungen über die galvanische Kette*, Leipzig: Brockhaus, 1831. Fechner, Gustav Theodor, "Ueber die Verknüpfung der Faraday'schen Inductions-Erscheinungen mit den Ampère'schen elektro-dynamischen Erscheinungen," *Annalen der Physik und Chemie* 64, 1845: 337-345, on p. 345.

that most captured his attention. He advocated a form of monism called psychophysical parallelism, arguing that psychical and physical states—indeed, psychical and physical worlds—are two aspects of one reality and that their relation can be studied quantitatively. This led him, building on the work of Weber's brother Ernst Heinrich Weber, to the so-called Weber-Fechner law, relating the physical strength of a stimulus to its perceived psychical intensity.

With respect to Weber's *Grundgesetz*, Fechner's *Ueber die physikalische und philosophische Atomenlehre* of 1855 is his most important work. [42] In this wide-ranging polemical tract, Fechner aimed to counter the currently dominant anti-atomism among German philosophers (as opposed to physicists). Ever since Kant's *Metaphysical Foundations of Natural Science* a number of philosophers had been pursuing forms of dynamism, meaning the view that the ordinary matter of our experience is constructed in the dialectics of nature from an underlying continuum of forces. "According to most dynamicists, a conflict of opposing forces is supposed to be what makes a body out of force." Two of Fechner's targets were Schelling and Hegel in their pursuit of the absolute or *Ding an sich*, but Herbart came in for special critique because his purely metaphysical monadology could look similar to the physical atomism that Fechner himself defended. [43]

For Fechner the real world was a world of *sinnliche Erscheinungen* (sensory appearances, or phenomena) and any idea of a *Ding an sich* behind appearances was pure fantasy. Such appearances were epitomized by what could be directly touched or grasped, but they extended much further. "If one asks in general what the world consists of in the last instance, then it is *Erscheinung* (*Selbsterscheinung* in mind and God, objective *Erscheinung* in nature): laws of *Erscheinung*; determinations, connections, and relations of Erscheinungen; which include the possibility of forthcoming and new Erscheinungen. Otherwise there is nothing and behind them there is nothing." [44] Within this simultaneously realist and phenomenalist perspective Fechner presented his atomistic world view, arguing that atomism best represented the totality of empirical and mathematical appearances known to physicists and therefore had the most probable claim on reality. [45] In this effort he also relied on several

---

42 Fechner, Gustav Theodor, *Ueber die physikalische und philosophische Atomenlehre*, Leipzig: Mendelssohn, 1855.

43 Fechner, *Atomenlehre*, 107, 164.

44 Fechner, *Atomenlehre*, 94, see also 90-99, 113.

45 See Heidelberger, *Nature from Within*, 137-154.

of the tools of believability that Johnston ascribes to the narration of Greek myths.

Conceptual metaphor. Under the conceptual framework of atomism Fechner sought to integrate a wide diversity of phenomena in the physical world. By atoms he understood discrete, indestructible atoms, *Grundatome* or *letzten Atome*, with forces acting directly at a distance between them. And citing Weber, along with prominent French physicists (Moigno, Séquin, Cauchy, Ampère), he adopted the view that these atoms could best be considered as unextended point atoms.[46] Crucially, the forces were nothing in themselves; they could not be thought of independent of the atoms; nor did they inhere in or emanate from individual atoms; so one atom could not be said to act on another. It was only "the category of *Zusammensein* [being together, or interrelation] that defined the concept of force, not an inner essence of matter." Or again, "The concept of force ... is a relational concept, which has meaning only for the *Zusammensein* of matter."[47] Forces were relations in space and time between atoms, which physicists knew only as laws. Thus Fechner's basic physical image was of a *pair* of point atoms moving with respect to one another and expressing in their relation the law of force that governed their relative motion. With this concept of action at a distance between atoms Fechner sought to open up the unobservable world to physical understanding grounded in *sinnliche Erscheinungen*. "Atomism is at once the key with which the physicist unlocks the door of a room closed to the senses and opens up its connection with what is immediately accessible to him."[48]

Serial narration. By the time Fechner's *Atomenlehre* appeared in 1855 he had been publishing articles and books that concerned atomism for thirty years. In this sense the *Atomenlehre* had a serial character, although that background appeared only occasionally in the text. More interesting is what might be thought of as the historical seriality of other physicists, mostly French, on whom Fechner depended. He had only to mention their names at critical junctures, for they were well known to all physical scientists. The series of their works portrayed a continuing French pursuit of action at a distance between "material points." Its coherent development, amidst lively debate, began perhaps from Laplace's popular reworking of Newtonian universal gravitation in his *System of the World* (1796) and in his five-volume mathematical treatise on

---

46   Fechner, *Atomenlehre*, 73, 79-81, 161-163.
47   Fechner, *Atomenlehre*, 109, 112.
48   Fechner, *Atomenlehre*, 32.

Celestial Mechanics (1799-1825). It continued through Poisson's adaptation of the inverse square law to electric and magnetic fluids (1811, 1821); Fourier's analysis of heat conduction as radiation between molecules; Fresnel's theory of light as transverse waves in the ether (1822, originally much contested); and Cauchy's representation of this ether as an elastic medium consisting of imponderable atoms (1835-36). Included of course was Ampère's electrodynamics (1824), which culminated in Weber's *Grundgesetz*. Fechner himself had been especially active in bringing the French tradition to Germany, both in his extensive translations (sometimes amounting to full rewritings) of comprehensive textbooks by Jean-Baptiste Biot on physics (four volumes, 1824; five volumes, 1828-1829), Louis Jacques Thénard on chemistry (seven volumes, 1825), and in his own *Repertorium der Experimentalphysik* (three volumes, 1832).

To think of this sequence in terms of serial narration of an atomistic world view, rather than simply as a tradition, is to think of it as an ongoing saga with a continuing story line and with surprising new episodes at every turn. Many physicists had either lived through the series or followed it in retrospect, attentive to the controversies within it, with expectations for what would come next, and looking back to reinterpret earlier episodes, such as the wave theory of light after Cauchy.[49] Fechner exploited such episodes in familiar vignettes, reiterating for example how Poisson had been forced to change his views on the polarization of light. These are all aspects that contributed to the believability of atomism. From a rhetorical perspective it was particularly effective for Fechner to fashion his own narrative with the ever-present foil of the dynamicists to enliven it throughout.

Multipliers. Here seriality merges into other multipliers of believability, such as multiple narrators who only partially agree. For example, Fechner could use the French series to enhance the credibility of his atomism despite the fact that in detail it presented a contrasting conception of his basic conceptual metaphor. While Fechner and Weber considered force as a shorthand for the interrelation of a pair of atoms, their *Zusammensein*, the French spoke of force as emanating from one atom and acting on another. The distinction is striking in the case of Gauss and Weber, who worked closely together at Göttingen. In a long article on inverse square forces, Gauss followed the French in writing of "a material point out of which a repulsive or attractive force acts."[50]

---

49   Fechner, Atomenlehre, 18.
50   Gauss, Carl Friedrich, "Allgemeine Lehrsätze in Beziehung auf die im verkehrten Verhältnisse des Quadrats der Entfernung wirkenden Anziehungs-und Abstossungs-Kräfte." In Resultate aus den Beobachtungen des magnetischen Vereins im Jahre 1839,

Figure 6: Visual depictions of (a) Gauss's mode of representing the force at a point p emanating from an electric mass point e and (b) Weber's comparable representation of the force between e and e' as an abstract relation in space.

Expressed mathematically (and visually in figure 6a) this meant that he calculated the force at an empty point of space produced by the material point (i.e., at point p the force F_p of an atom e at a distance r would be $F_p = e/r^2$, or the force per unit mass that would be exerted on another atom *if it were placed there*). In contrast (figure 6b) Weber expressed the force as a relation between a pair of atoms e and e', $F_p = ee'/r^2$. Ironically, Hermann Helmholtz, in formulating his classic work on energy conservation in 1847, used the Fechner-Weber conception of force in terms of atom pairs even while citing Gauss.[51]

A similar multiplicity of voices continued their expression in the period following Fechner's *Atomenlehre* with its reliance on Weber's *Grundgesetz* as its epitomy. Helmholtz criticized the law for its time dependence, which he thought violated conservation. This produced a long and sometimes acrimonious dispute with Rudolph Clausius and Weber, who showed that it did not.[52]

edited by C. F. Gauss and W. Weber, Leipzig, 1840. Reprinted in Gauss, Carl Friedrich, Werke, Göttingen: Königlichen Gesellschaft der Wissenschaften, 1877, vol. 5, 195–242, on p. 198-201.

51   Wise, M. Norton, Aesthetics, Industry, and Science: Hermann von Helmholtz and the Berlin Physical Society, Chicago: University of Chicago Press, 2018, 271.

52   Bevilacqua, Fabio, "Theoretical and Mathematical Interpretations of Energy Conservation: The Helmholtz-Clausius Debate on Central Forces 1852-54." In Universalgenie Helmholtz, 89-106. Darrigol , "Helmholtz's Electrodynamics and the Comprehensibility of Nature." In Universalgenie Helmholtz, 216-242.

A full telling of this controversy would involve a number of other major actors and their commitments. I emphasize here only that the controversy provided a powerful multiplier for belief in atomism and for Weber's *Grundgesetz*, even as Maxwell's electromagnetic field theory became a prime competitor.

Story World. In its most general form the world that Fechner presented to his readers was a world of discrete things within which he aimed to join all of the physical sciences in a common structure. If gravitation and electricity provided the groundwork of atom-pairs and inverse square forces to which all else would ultimately be reduced, he came to this position within a much broader vision of an atomic system as analogous to a planetary system, a Laplacian system of the world, extending from the stars moving in the heavens to the planets of the solar system to atomic systems making up the molecules of ponderable matter and those of the imponderable ether. Under this universal scheme of discreteness and systems all of the subjects of the physical sciences had already made great progress: light, heat, elasticity, cohesion, chemical combination, crystallography, etc. "Thus through atomism everything from the largest to the smallest and in the most diverse directions is encompassed within a single realm, and a general clarity runs through this realm."[53]

Within this material world of unifying clarity, Fechner had also to make room for contemplation of the "highest and final things," of God, morality, freedom, life and death. The dynamicists supposed that a world conceived as a continuum of forces was more suited to relating matter and spirit than a world of atoms, which he firmly denied. "The same spirit that runs through atomism must be conceivable as a whiff of the same spirit that runs through heaven itself, whether it can exist with God or God with it." The atomistic world in fact supplied an illuminating image of a social organization based on the "principle of individuality" and spiritual freedom rather than of everyone tied to their neighbor without independence. In short, "an atomistic world is a structure worthier of the most exalted idea of God and indescribably more beautiful than the dynamical."[54] Here was a story world into which Fechner hoped his audience could project their most wide-ranging beliefs, or at least suspend their disbelief in atomism.

Pragmatic effect. It was the molecular structure of matter that Fechner particularly exploited to make the superiority of atomism seem almost accessi-

---

53   Fechner, Atomenlehre, 36.
54   Fechner, *Atomenlehre*, 119, 122.

ble to the senses. For example, if molecules consisted of atomic systems that could take different arrangements, then phenomena like isometry, in which substances with the same chemical composition have different properties, became intuitively realistic, making "the advantage of the atomistic conception palpable [*fühlbar*] for the unprejudiced."[55] That was already a major contribution to suspension of disbelief. But it also sharpened the further question of how atomic systems could actually be structured as stable molecules by forces between atoms.

For this question Fechner appended to his more evidentiary text a speculative chapter containing a "Hypothesis on the General Force-law of Nature." Here he relied on the credibility of Weber's earlier suggestions for multi-body forces and time-dependent forces to unfold a much more expansive view. If gravitational and electrical forces expressed the relation of two particles, why suppose that nature would have stopped there? "Is it not possible that results appear here that depend on forces that are determined jointly by the *Zusammensein* of more than two particles?"[56]

Figure 7: A representation of Fechner's conception of an irreducible multi-body force as the Zusammensein of five particles.

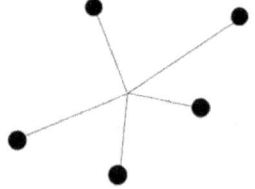

On this basis (figure 7) Fechner proposed an ascending series of forces as the number of particles in a system increased and whose strength decreased increasingly rapidly with distance between the particles. These higher-order forces would be unobserveable at large distances but would gradually

---

55   Fechner, *Atomenlehre*, 37.
56   Fechner, *Atomenlehre*, 184.

come into play as more particles at smaller distances made up more complex molecules. Briefly put, "In every combination of arbitrarily many particles there rules a force, whose strength and direction [attractive or repulsive] are determined by the interrelations of the *Zusammensein* of all the particles at once ...."[57] This conception might extend all the way from chemical elements as systems of *Grundatome* to a force governing the totality of the parts of an organism, which would encompass within it many subordinate systems and their forces.

Fechner would also have liked to be able to understand all of the phenomena ascribed to imponderable substances, such as light, electricity, and magnetism, in terms of the same *Grundatome* that made up ponderable matter, while referring them to the oscillations of individual atoms rather than to atomic systems and higher order combinations of molecules. But too little as yet was known about them. He could, however, suggest that Weber's velocity and acceleration dependent *Grundgesetz* for electric masses would very probably need to be extended to the atoms of normal matter. That would explain such things as the expansion of bodies by heating, which would give their particles a greater velocity and perhaps therefore a weaker attractive force between them.[58]

It is apparent that in this last chapter Fechner was reaching for a pragmatic effect, that having already found his atomistic world believable his audience would be open to a wide range of possible realties that might well fall within that general conception. If so, dynamism had been defeated by the rhetorical techniques of the *Atomenlehre*.

3. Conclusion

I have attempted to show three things: (1) how Maxwell and Weber structured their pictures of electromagnetic action as sophisticated narratives that integrated diverse aspects of the subject; (2) how the writings of Faraday and Fechner placed the particular stories of Maxwell and Weber in a wider story world, which enhanced their believability; (3) how evoking this story world depended on the kind of narrative techniques that Johnston finds in her *Story*

---

57   Fechner, *Atomenlehre*, 193.
58   Fechner, *Atomenlehre*, 207.

of Myth. My question now concerns the implications of this reading for comparison.

Thomas Kuhn once wrote that "Theories, as the historian knows them, cannot be decomposed into constituent elements for purposes of direct comparison either with nature or with each other." He was writing here about the holistic character of what he famously called paradigms in science and the similarly holistic character of historical narratives about science. Both theories and narratives were like "pictures" or "patterns." The historian's job was to construct "a plausible narrative involving recognizable motives and behaviors" that fit into a coherent pattern.[59] Paul Roth has discussed this perspective with reference to how Kuhn drew on the philosopher of history Louis Mink and his concept of "synoptic judgement" in historical narratives. "The distinctive characteristic of historical understanding," Mink argued, "consists of comprehending a complex event by 'seeing things together' in a total and synoptic judgement which cannot be replaced by any analytic technique."[60]

Maxwell seems to have intended something similar when he wrote that the aim of his physical analogy of lines of force as flow lines had been "to present the mathematical ideas in an embodied form ... and not as mere symbols, which neither convey the same ideas, nor adapt themselves to the phenomena to be explained." Not that the symbolic representation would be wrong but that it would be too thin; it would not evoke the full depth of mental images and bodily sensations of the embodied analogy, marked by its vividness and fertility. Fechner made a related point in his presentation of atomism in terms of sensory appearances: "through their conception we better orient ourselves in the visible and palpable."[61] Both Maxwell and Fechner, in their very different ways, sought to arouse the creative imagination through their use of narrative techniques, with their power to make fictional entities into realistic possibilities.

---

59   Kuhn, Thomas S., "The Relations between the History and the Philosophy of Science." In The Essential Tension: Selected Studies in Scientific Tradition and Change, Chicago: University of Chicago Press, 1977, 3-20, on p. 19, 17. Rather: 17, 19 ?

60   Roth, Paul A., "The Silence of the Norms: The Missing Historiography of The Structure of Scientific Revolutions," Studies in History and Philosophy of Science 44, 2013: 545-552, on p. 550-551. Mink, Louis, "The Autonomy of Historical Understanding," History and Theory 5, 1966: 24-47, reprint in Mink, Historical Understanding, edited by B. Fay, I. O. Golob, and R. T. Vann, Ithaca and London: Cornell University Press, 1987, 61-88, on p. 82.

61   Fechner, Atomenlehre, 105.

It has been notoriously difficult for historians and philosophers of science to give a clear articulation of what exactly the something extra is that goes beyond the component parts of holistic entities. The long-standing tradition of treating narrative and natural science as dichotomous has not helped, most famously in Carl Hempel's argument that only the natural sciences in their lawlike, deductive form could provide explanations.[62] But the natural sciences themselves, in their now so pervasive studies of nonlinear dynamical systems, have found it necessary to employ holistic concepts of complexity, emergence, entanglement, order out of chaos, and embodiment that belie any easy distinction between narrative and natural science. They have also helped to stimulate new forms of historical analysis.[63] A closely related result among historians of science has been a growing emphasis on the functions of narrative within the sciences themselves. I have argued elsewhere, for example, that the widespread use of model-based simulations to understand complex processes often takes the history-like form of following out the possible developmental narratives generated by the (fictional) model. These explorations sometimes include a key role for representation of the simulations as movies, or visual narratives.[64] Such visualizations take to a literal level the idea of a historical narrative as being like a picture or pattern.

An even more general approach to the role of narrative knowing in complex domains, particularly in the social sciences, has been pursued by Mary Morgan, who (like Kuhn, and citing Mink) emphasizes the coherence-making power of narratives, their capacity to order and to fit together in a coherent pattern a variety of disparate elements that otherwise would not seem to belong together. This integrating capacity is very much in evidence in the narratives of electromagnetism that I have described. As Faraday put it to Ampère, lacking the capacity for abstract synthesis, "I am obliged to feel my way by facts closely placed together," by their "connexity," as one interpreter puts

---

62   Hempel, Carl G., "The Function of General Laws in History." [1942] In Aspects of Scientific Explanation, and Other essays in the Philosophy of Science, edited by C. G. Hempel, London: Macmillan, 1965, 232-243.

63   Stark, Laura, "Emergence," in Focus section on Explanation, Isis, 110, 2019: 332-336, discusses the import of this movement, with key references.

64   Wise, M. Norton, "On the Narrative Form of Simulations," Studies in History and Philosophy of Science 62, 2017: 74-85 (special issue on narrative science, edited by M. S. Morgan and M. N. Wise).

it.[65] Morgan uses a visual analogy from a painting by Peter Breugel, which depicts numerous small groups of children engaged in seemingly unrelated activities. Properly ordered, however, they fit together under the higher-order concept and title of "Children's Games." Interpreting more contextually, the whole ensemble represents a moral story of how in the eyes of God, people are like children. Morgan captures the action of ordering, relating, and knitting together in the term "colligation."[66]

Interestingly, it is the same word that Jouni-Matti Kuukkanen uses to capture the "essence" of narrativism in historiography: "narratives ... are colligatory additions to our understanding of the past." He has developed this view at length in his *Postnarrativist Philosophy of Historiography*, where colligatory concepts provide the centerpiece of his argument.[67] Johnston's conceptual metaphors serve a similar purpose.

All of these examples have a common theme, which I fully endorse. Many works of both history and science—especially when dealing with complexity—can be fruitfully analyzed in terms of narrative. The narrative reading suggests that understanding accounts of particular phenomena requires that they be treated holistically, attending to the way in which they incorporate

---

65   Cited by Darrigol, Electrodynamics from Ampère to Einstein, 21, using the term "connexity." Faraday to Ampère, 3 September 1822, in James, Frank, editor, The Correspondence of Michael Faraday, vol. 1: 1811-1831, London: The Institution of Engineering and Technology, 1991.

66   Morgan, Mary S. "Narrative Ordering and Explanation." Studies in History and Philosophy of Science 62, 2017: 86-97, on p. 88-89 (special issue on narrative science, edited by M. S. Morgan and M. N. Wise). Morgan actually prefers Mink's later discussion of "configuring" to "synoptic judgement," because it emphasizes the active process of analysis that leads to "colligation" as a result and to another important mode of ordering by "juxtaposition," which highlights the "puzzles" within a narrative whose resolution yields deeper understanding (pp. 90-93).

67   Kuukkanen, Jouni-Matti, "The Missing Narrativist Turn in the Historiography of Science," History and Theory 51, 2012, 340-363, on p. 357. Kuukkanen, Jouni-Matti, Postnarrativist Philosophy of Historiography, Basingstoke, UK: Palgrave Macmillan, 2015, 97-130. While giving pride of place to colligation, Kuukkanen rejects two other tenets of narrativism in historiography, holism (especially for texts, but perhaps not for "holistic" colligatory concepts, p. 112) and representationalism (on the analogy with visual art and representations as pictures), both of which are characteristic of my own treatments of narrative in the physical sciences. But I suspect that his rejection of these terms results from overly strict definitions, which would not apply to the highly visual and holistic cases of simulation that I have analyzed. The issue deserves much more discussion than I can offer here.

their diverse strands into a discursively elaborated conception of a portion of the world that coheres together. It is the construction of this coherence that has led me to treat Maxwell's and Weber's essays in terms of the narrative unfolding of images of electromagnetic action, depictions designed to make realistic fictions plausible or believable.

The two approaches of Maxwell and Weber are so radically different, however, that they appear to have belonged to different conceptual worlds, with very little overlap between them. This has motivated my consideration of how their believability depended in part on their being located in different story worlds—represented by Faraday and Fechner—that extended well beyond their particular conceptual constructions and that made them seem familiar. That suggestion gains weight from the analogy with the believability of Greek myths. Johnston has argued that a major problem in the treatment of myths in classical scholarship has been their abstraction from the actual cultural and social life of the Greeks, which has entailed removal of individual myths from the story world and the narrative practices in which they were embedded. This sort of abstraction has made it difficult to understand why (or even that) the myths were believable. Her argument is that gods and heroes were believable to the Greeks because they were taken as part of the real world, either at present or in the human past, and thus seemed part of the normal world of human action. Narrative techniques and practices performed this familiarizing role by blurring the lines between known realities and fictional possibilities. Something similar, I am proposing, operated with respect to both Maxwell's and Weber's accounts of electromagnetism.

I have so far left open the question of how two narratives that seem to occupy different worlds can be compared. That question might seem to raise the fraught issues of Kuhnian incommensurability of paradigms, whereby Maxwell and Weber simply could not understand each other and comparison was impossible. Like most other historians and philosophers of science, I do not find this view tenable in any strict sense. But appreciation for, and a willingness to entertain alternative possibilities or competing views, is a different matter. Here is where treating scientific texts as holistic narratives occupying different story worlds is worthwhile. Without in any way compromising an appeal to empirical adequacy, mathematical unity, and comprehensiveness, it immediately raises the question of how effectively narrated the two representations are, and that is a question not only of their own narrative virtues but also of their being situated within a broader story world capable of enhancing their credibility. The analogy with Greek myths has suggested several aspects of effective narration that should be important: conceptual metaphor, serial narration, multipliers, and story world. Evaluation of their effectiveness will of course be a subjective matter and will involve judgements of such things as heuristic power, aesthetic appeal, emotional grip, and philosophical prefer-ence. This does not make everything arbitrary or equal but it does imply that comparison of competing accounts will require the kind of holistic judgement that we expect of narratives and that is well captured by colligation.[68]

That returns me to my starting point and to Maxwell's question about why we should entertain alternative possible realities. He did all he could to pro-vide motivation for "imagining" lines of force and an electro-tonic state occu-pying every point of space when Weber had already given a perfectly compre-hensible depiction of time dependent forces acting immediately at a distance. He did not attempt to argue on purely rational grounds that his view was preferable, but only that it would be preferable to many minds who found the conceptual and sensory immediacy of physical analogies more satisfying than abstract mathematical formulas. And he was fully aware that others would differ about which was more satisfying. For the moment, therefore, until fur-ther empirical or theoretical developments were available to support one or the other perspective, he could only remark that "it is a good thing to have two ways of looking at a subject, and to admit that there *are* two ways of looking at it." Perhaps that is a key lesson of the narrative reading of scientific works. It draws out their power to produce vivid synthetic depictions that capture the creative imagination while also making it apparent that comparisons will involve the same kinds of valuations that are familiar for literary works and works of art.

3.1   A Comment on Temporality

I approach the question of whether the term narrative necessarily implies a *temporal* sequence of connected events from a historian's perspective. Many historical works are of course devoted to temporal dynamics and philosophers of history coming from a phenomenological perspective, such as Paul Ricoeur in *Time and Narrative*, take the lived experience of time to be fundamental to

---

68   Kuukkanen, Postnarrativist Philosophy of Historiography, 123-128, does not include such subjective evaluations of narratives but limits himself to a set of epistemic values for their colligatory concepts: exemplification, coherence, comprehensiveness, scope, and originality.

human understanding, and thus to history. (As noted above, I have adapted this view for the way in which simulations provide understanding of physical processes.) But much historical writing is not focused on temporality. An example is Carl Schorske's *Fin de Siecle Vienna*, which is concerned rather with providing a vivid depiction of a memorable cultural constellation than with analyzing its rise and fall. More generally, historians like other social scientists are often more concerned with understanding and depicting the structure of relations characteristic of a particular culture or situation than with tracing or accounting for the temporal course of its development, though both are often in play. This preference can extend even to an antipathy for the focus on time. Louis Mink is perhaps the most famous representative, arguing that we can understand a narrative, even a temporal narrative, only retrospectively, for it is only in retrospect that we can obtain the synoptic judgement mentioned above. "In the understanding of a narrative the thought of temporal succession vanishes" so that "time is not of the essence of narratives."[69]

Surely this is too extreme, but it does suggest that the power of narratives in general can be better characterized by their ability to draw things together in a conceptual scheme, their capacity for colligation, as Morgan and Kuukkanen would have it, than by their temporality per se. While many narratives will depend on temporal ordering to attain their colligatory concepts, and even on the experience of following a process in time to gain understanding, many others will not, or they will use both temporal and non-temporal descriptions in a complementary fashion. For example, Morgan stresses the puzzle-raising functions of Clifford Geertz's classic account of Balinese cockfighting while Kuukkanen focuses on the argumentative character of Christopher Clark's depiction of events leading up to WWI in *the Sleepwalkers*.[70] From

69   Mink, Louis, "History and Fiction as Modes of Comprehension," New Literary History 1, 1970: 554-555, reprint in Mink, Historical Understanding, edited by B. Fay, I. O. Golob, and R. T. Vann, Ithaca and London: Cornell University Press, 1987, 56-57.
70   Morgan, "Narrative Ordering and Explanation," 92-93. Geertz, Clifford, "Deep Play: Notes on the Balinese Cockfight," Daedalus 101, 1972: 1-37. Clark, Christopher, The Sleepwalkers: How Europe Went to War in 1914, London: Penguin, 2012. Kuukkanen, Postnarrativist Philosophy of Historiography, 92-96. Kuukkanen nevertheless speaks of the temporal part of a historical text as the "narrative" part, or "narrativity," (also 73-75) but that appears to play no fundamental role in his important analysis of colligatory concepts.

this perspective, temporal ordering figures as a (critically important) subset of narrative ordering.[71]

71   I thank Mary Morgan for her valuable comments on this paper.

Historical Narrative versus Comparative Description?
Genre and Knowledge in Alexander von Humboldt's *Personal Narrative*

Christine Peters

Alexander von Humboldt's *Personal Narrative of Travels to the Equinoctial Regions of the New Continent* (1814–1825)[1] appears as an especially productive source of research on the relationship between narrating and comparing in the sciences. Humboldt frequently shifts from what he calls a "historical narrative," meaning the recounting of episodes in his voyages, to a comparative "description" of individual objects. These shifts between two different forms of scientific travel writing do not just constitute the structure of Humboldt's travelogue. They also define large parts of the genre of travel writing in the eighteenth and nineteenth centuries. What appears unconventional about Humboldt's travelogue is not the hybrid structure itself but the immense effort he makes in commenting on this structure and in evaluating the relationship between historical narrative and description. He frequently stages the two as opposed modes of travel writing and even as mutually exclusive means of structuring a travelogue. In the descriptive sections, Humboldt engages more frequently, one might even say excessively, in comparing. In many cases, as in, for example, his description of Tenerife, which represents a central object of investigation in this paper, he even explicitly defines "comparison"[2]

---

1   Originally, the Personal Narrative was published in French as the Relation Historique (1814–1825), but it was translated almost simultaneously into German and English. This paper focuses on the English translation.

2   von Humboldt, Alexander, Personal Narrative of Travels to the Equinoctial Regions of the New Continent, during the Years 1799–1804, vol. 2, New York: Cambridge University Press, 2011 [1814], on p. 221.

or "comparing"[3] as one of the main purposes of these "descriptive" passages. Therefore, I argue that he does not contrast the historical narrative just with description but, more specifically, with *comparative* description.

This paper aims to historicize Humboldt's genre-specific notions of narrating and comparing, and to interpret them in a broader epistemological context. To do so, I firstly argue for the viability of a criteria-based but, at the same time, scalar definition of narrative as well as for a historically restricted approach to narrating and comparing—an approach that links narrative theory to recent praxeological research on comparison. I shall then investigate Humboldt's definitions of "historical narrative" and "description" in the context of nineteenth-century travel writing. Furthermore, I shall analyze the passages in which Humboldt shifts from one to the other and use these to outline the role of comparing in his descriptive text passages. I seek to show that Humboldt treats the "historical narrative" and the comparative "description" not only as opposed types of travel writing but also as conflicting forms of knowledge acquisition; and that Humboldt's preference for "comparative" writing promotes a trans-areal and universal epistemological project. My claim there is as follows: Humboldt seems to reject the historical narrative, because the singularity and particularity which is essential to narrative and its temporal sequentiality does not fit with his universal epistemological aim. I shall try to show how Humboldt partially resolves this contradiction when he turns toward the history of the Earth which he organizes as a sequence of geological events in order to uncover the universal laws underlying geological change.

Against this general background, I shall show how Humboldt actually deals with these seemingly contrasting modes of writing, focusing on acts of comparing in his descriptive text passages. I argue that these text passages partly contradict Humboldt's programmatic distinction of narrative, description, and comparison, and that comparing temporarily adopts qualities and functions of narrating, such as building temporal sequentiality or providing a viable point of view.

---

3   von Humboldt, Alexander, Personal Narrative of Travels to the Equinoctial Regions of the New Continent, during the Years 1799–1804, vol. 1, New York: Cambridge University Press, 2011 [1814], on p. 229.

1. Toward Scalar and Historically Restricted Definitions of Narrative

An inquiry into Humboldt's forms of narrating and comparing soon leads to the question regarding what these terms actually cover. The search for definitions becomes especially central in the field of narrative theory, or what is referred to more commonly as narratology in literary studies. Whereas scholars from literary and cultural studies have produced a wide range of definitions relying on various criteria such as sequentiality, a narrating instance, an event-like nature, temporality, causality, changeability, or experientiality,[4] scholars from the history of science have been reluctant to work with a criteria-based definition of narrative. As Christina Brandt points out, scholars from science studies mostly rely on a general notion of narrative, extending the term to a variety of possible meanings such as a textual structure, a text-independent structure, a scientific discourse, or a system of values.[5] Taking this variety of approaches into account, scholars might ask not only which criteria are needed to define a narrative but also whether criteria are needed at all. In the following passage, I shall briefly address the latter question using the example of cultural studies. I shall then propose a scalar definition of narrative that takes into account the historicity and changeability of both narrative and its definitions.

Even though scholars from cultural studies or "cultural analysis"[6] have approached narrative in order to address questions of cultural significance, they have relied heavily on what are originally structuralist categories from the

---

4   Bal, Mieke, Narratology: Introduction to the theory of narrative, 3rd ed., Toronto: University of Toronto Press, 2009. Fludernik, Monika, An introduction to narratology, London: Routledge, 2009; Martínez, Matías, editor, Handbuch Erzählliteratur: Theorie, Analyse, Geschichte, Stuttgart: Metzler, 2011. Martínez, Matías, editor, Erzählen: Ein interdisziplinäres Handbuch, Stuttgart: Metzler, 2017. Nünning, Ansgar, "Wie Erzählungen Kulturen erzeugen: Prämissen, Konzepte und Perspektiven für eine kulturwissenschaftliche Narratologie." In Kultur – Wissen – Narration: Perspektiven transdisziplinärer Erzählforschung für die Kulturwissenschaften, edited by A. Strohmaier, Bielefeld: transcript, 2013, 15–53. Ryan, Marie-Laure, "Toward a definition of narrative." In The Cambridge companion to narrative, edited by D. Herman, Cambridge: Cambridge University Press, 2007, 22–35. Schmid, Wolf, Elemente der Narratologie, 3rd ed., Berlin: De Gruyter, 2014.

5   Brandt, Christina, "Wissenschaft." In Erzählen: Ein interdisziplinäres Handbuch, edited by M. Martínez, Stuttgart: Metzler, 2017, 211f.

6   Bal, Narratology, on p. 225.

field of literary studies in order to define and analyze it. Especially the categories *focalization* and *voice* have frequently been used to explore cultural and political issues such as questions of representation and accessibility[7] or the narrative construction of cultural identities and worlds.[8] The main advantage of these interdisciplinary approaches is, firstly, that narrative becomes the subject of cultural studies, and, secondly, that the formal aspects constitutive for narrativity are valued as carriers of meaning themselves—and not only as simple technical stuff.[9] I argue that, in an analogous manner, scholarly fields such as the history of science, science studies, or historical epistemology could gain from an approach which implies not only the interpretation of certain narrative texts but also the investigation of their specific narrative qualities and epistemological functions. In Humboldt's case, so the central claim of this paper, reflections on narration (and on comparison) as well as the actual way of narrating (and comparing), are closely intertwined with the epistemological claims of the text. This connection between the specific narrative setup of a text and its epistemological implications is not exclusive to Humboldt's scientific travel writing. It appears to be rather characteristic for articles in the life sciences during the nineteenth century where it is also combined with extensive descriptive passages. Martina King's attempt to combine historical narratology and the history of the life sciences in the early nineteenth century shows that a processual understanding of nature is especially suited for sequential representation, and that a specific type of narrator—namely, a homodiegetic narrator who is part of the world he narrates—guarantees epistemic authenticity. Whereas narrative sequencing in itself reduces contingencies and instead brings about causality and coherence, a homodiegetic narrator furthermore authenticates knowledge by serving as a witness and by embedding his perceptions in a specific temporal and spatial setting.[10] The evidence King proposes strongly suggests that the narrative setup of scientific writing in the early nineteenth century carries epistemological meaning.

---

7   Bal, Mieke, "Intercultural Story-Telling." In Kultur – Wissen – Narration: Perspektiven transdisziplinärer Erzählforschung für die Kulturwissenschaften, edited by A. Strohmaier, Bielefeld: transcript, 2013, 289–305.

8   Nünning, "Wie Erzählungen Kulturen erzeugen," 15–53.

9   Nünning, "Wie Erzählungen Kulturen erzeugen," 27–31.

10   King, Martina, "'Ich habe im Sommer des Jahres 1838 eine Reihe von Beobachtungen angestellt': Naturwissenschaftliches Erzählen im frühen 19. Jahrhundert," DIEGESIS 6(1), 2017: 20–45.

However, criteria-based definitions are usually developed in literary studies, which is why they refer mostly to fictional narratives. Therefore, they do not necessarily provide a sufficient set of criteria to deal with the specific narratological setup of factual texts such as the scientific travelogue investigated in this paper. Recent research on the matter shows that factual narration is still a rather marginal subject in literary studies, even though scholars frequently stress the omnipresence and anthropological importance of narrative beyond literary prose.[11] Not only Martínez and Klein but also Fludernik, Falkenhayner, and Steiner distinguish between fictional and factual narrative, in a way that explains why the two forms of narration function differently. Whereas Martínez and Klein argue that factual narration is characterized mainly by its referential truth claims (factual narration refers to real events or to reality in general),[12] Fludernik uses a number of criteria to outline the differences between the two such as the possibility (or impossibility) of collective or nonanthropomorphic actors, the extent and type of emotional immersion, the differentiation (or nondifferentiation) of author and narrator, the level of experientiality, the quantity and importance of argumentative text passages, or the possibility (or impossibility) of unreliable narration.[13] Criteria-based definitions of narrative in literary studies do not usually concern themselves with these differentiations that can, however, be central to the investigation of scientific narration. Instead, they focus more on criteria that enable scholars to deal with the wide range of narratological phenomena in fiction. But, as the next example will show, the question of whether or not to rely on a fixed set of criteria tackles both heuristic usefulness and the building of one's corpus. A definition of narrative based on strict criteria might exclude sources that scholars aim to investigate and tentatively classify as narrative.

---

11   Fludernik, Monika, Nicole Falkenhayner, and Julia Steiner, "Einleitung." In Faktuales und Fiktionales Erzählen: Interdisziplinäre Perspektiven, edited by M. Fludernik, N. Falkenhayner, and J. Steiner, Würzburg: Ergon, 2015, 7–22. Martínez, Matías, and Christian Klein, "Wirklichkeitserzählungen: Felder, Formen und Funktionen nicht-literarischen Erzählens." In Wirklichkeitserzählungen: Felder, Formen und Funktionen nichtliterarischen Erzählens, edited by M. Martínez and C. Klein, Stuttgart: Metzler, 2009, 1–13.

12   Martínez and Klein, "Wirklichkeitserzählungen," 6.

13   Fludernik, Monika, "Narratologische Probleme des faktualen Erzählens." In Faktuales und Fiktionales Erzählen: Interdisziplinäre Perspektiven, edited by M. Fludernik, N. Falkenhayner, and J. Steiner, Würzburg: Ergon, 2015, 116ff.

Binary definitions such as the ones proposed by Matías Martínez and Michael Scheffel have proven highly productive when applied as tools for textual analysis but might be interpreted as overly exclusive when it comes to questions of classification.[14] For example, in his latest publication on this issue, Martínez defines narrating as a representation of events that refers to a specific subject; is organized in a temporal sequentiality; and shows spatial, temporal, and causal contiguity. Apart from these three mandatory criteria, Martínez refers to a large number of optional criteria such as a double temporality, mediation through a narrator, causality, human intentionality, completion, event-like nature, experientiality, tellability, or conversational necessities; and he argues that at least one of these optional criteria has to be met in addition to the three mandatory ones in order to speak of narrating.[15]

Applying this definition to Humboldt's *Personal Narrative* and to factual narratives in general reveals both the analytical usefulness and the exclusive potential of criteria-based definitions. This paper actually refers to some of the suggested criteria, mainly temporal sequentiality and mediation through a narrator, to describe the narrative setup of Humboldt's *Personal Narrative* and the universal and relational epistemological claims implied by it. However, if we want to go beyond the heuristic usefulness of individual criteria and question the general usefulness of criteria-based definitions for the analysis of factual narration, definitions such as the one suggested by Martínez appear to be too restrictive because of their binary structure. Holding a certain number of criteria as necessary increases the chances of excluding entire genres from narratological analysis, especially when investigating factual texts whose narratological setup is usually far less complex than that of fictional texts.[16]

---

14   Martínez, Handbuch Erzählliteratur. Martínez, Erzählen. Martínez, Matías, and Michael Scheffel, Einführung in die Erzähltheorie, 9th ed., München: C.H. Beck, 2012.

15   Martínez, Matías, "Was ist Erzählen." In Erzählen: Ein interdisziplinäres Handbuch, edited by M. Martínez, Stuttgart: Metzler, 2017, 2–5. For another influential definition of narrative, see Fludernik, Introduction to narratology, 6. Many of the criteria Fludernik incorporates in her definition resemble those suggested by Martínez but are organized in a different hierarchy. For example, whereas Martínez places "experientiality" in a long list of optional criteria, Fludernik stresses its central importance for the constitution of narrative. For a less extensive but equally restrictive binary definition, see Schmid, Elemente der Narratologie, 3f.

16   Fludernik, "Narratologische Probleme des faktualen Erzählens," 133.

To enable a more flexible analysis of narrativity and its connection to epistemological processes, I suggest "a scalar conception of narrative" as proposed by Marie-Laure Ryan. Some of the criteria she lists, such as causality or temporality,[17] resemble Martínez' criteria, but Ryan abandons the binary structure of mandatory and optional criteria:

> Rather than regarding narrativity as a strictly binary feature, that is, as a property that a given text either has or doesn't have, the definition proposed below presents narrative texts as a fuzzy set allowing variable degrees of membership, but centered on prototypical cases that everybody recognizes as stories. In a scalar conception of narrative, definition becomes an open series of concentric circles which spell increasingly narrow conditions and which presuppose previously stated items, as we move from the outer to the inner circle, and from the marginal cases to the prototypes.[18]

Martínez' list of criteria is far more extensive than Ryan's, and can therefore be considered a valuable tool box for analyzing all sorts of narratives. However, Ryan's scalar approach broadens the range of possible objects of investigation, encouraging analysis at the margins of narrativity and especially factual narrativity.[19] It also encourages analyzing the specific narrative qualities of a text instead of debating the question whether this text or the genre it represents can be regarded as narrative in the first place.

As a last preliminary note to my analysis of Humboldt's *Personal Narrative*, I aim to connect these narratological suggestions to recent research on comparison. This paper refers to a praxeological analysis of comparing that focuses less on comparison as a logical operation but more on what actors *do* when they compare.[20] Even though I do not focus on the materiality of Humboldt's practices of comparing, as a full-fledged praxeological approach

---

17   Ryan, "Toward a definition of narrative," 29.
18   Ryan, "Toward a definition of narrative," on p. 28.
19   Fludernik suggests a scalar or gradual approach to narrativity on yet another level: Factual and fictional narrating cannot be differentiated in a purely binary way either. Differences between the two forms of narrating are usually rather gradual. See Fludernik, "Narratologische Probleme des faktualen Erzählens," 118f. Schmid proposed a scalar approach concerning the event-like nature of narratives, arguing that not all events show the same level of "eventfulness." See Schmid, Elemente der Narratologie, 14ff.
20   Epple, Angelika, and Walter Erhart, "Die Welt beobachten und verändern: Praktiken des Vergleichens." In Die Welt beobachten und verändern: Praktiken des Vergleichens, edited by A. Epple and W. Erhart, Frankfurt a. M.: Campus, 2015, 7–31.

would require, I refer to praxeology: This analysis focuses on *practices* of comparing that are neither exclusive to one writer or text nor unchangeable in history. The investigation of narrative in the historical sciences could potentially gain from a praxeological perspective since it would shift attention from the analysis of *narrative* to the analysis of *narrating*. Just as Nicholas D. Paige has proposed for fiction—namely, "that along with asking what fiction 'is,' we might also ask if fiction always is, in the same way"[21]—I propose the following for narrative: Along with asking what narrative "is," we might also ask if narrative "*always* is, in the same way." This line of inquiry takes the historicity and changeability of forms of narrating (and comparing) into account, while simultaneously opening up the topic for a number of new questions: Why and how do forms or types of narrating and comparing change over time? Why are some forms more advantageous to writers at a certain point in history? Or, why do writers *perceive* or *present* some forms as being more advantageous than others?

Asking these questions changes the status of the definitions discussed above: Definitions of narrating and comparing can never be absolute or timeless. What actors do when they narrate or compare might gradually change over time. The criteria we work with are always preliminary. We might start analyzing a historic corpus with a certain set of criteria and then discover that the types of narrating or comparing at this specific point or period in history do not fully correspond with the definitions we have brought with us. We need definitions to identify acts of narrating or comparing in the first place, but these definitions must constantly be open to revision in light of the historicity and changeability of literary forms of writing and the epistemological implications that accompany them. This historical approach to narrative corresponds partly with recent research in the field of historical narratology. Even though much of narrative theory, especially within cultural studies, has stressed the omnipresence of narrative and its anthropological or cultural anchoring,[22] the historicity and evolution of literary modes of writing has been investigated occasionally by scholars on a grand scale; for example, in the recently published volume *Die Erzählung der Aufklärung* by Frauke Berndt

21   Paige, Nicholas D., Before Fiction: The Ancien Regime of the Novel, Philadelphia: University of Pennsylvania Press, 2011, on p. 2.

22   See, for example, Bal, Narratology. Bal, "Intercultural Story-Telling," 289–305. Fludernik, Introduction to narratology. Koschorke, Albrecht, Wahrheit und Erfindung: Grundzüge einer Allgemeinen Erzähltheorie, Frankfurt a. M.: Fischer, 2012.

and Daniel Fulda.[23] However, this approach still has to be applied to historical types of factual narration such as scientific narratives. This paper aims to contribute to this underestimated field of research by investigating some narratological particularities in Humboldt's *Personal Narrative*.

2.　"Historical Narrative" versus "Description"

Early in his travelogue, Humboldt aims to distinguish between narrative and descriptive travel writing:

> I had left Europe with the firm intention of not writing what is usually called the historical narrative of a journey, but to publish the fruit of my inquiries in works merely descriptive; and I had arranged the facts, not in the order in which they successively presented themselves, but according to the relation they bore to each other.[24]

This passage first introduces the "historical narrative" as a generic term, a term for "what is usually called the historical narrative."[25] However, Humboldt is not referring to narrating in general but to a specific narrative form of presenting a voyage's results to the general public. He characterizes this specific narrative form mainly by contrasting it with another form of scientific travel writing: namely, "descriptive" writing. Humboldt does not introduce the two as equal options but emphatically subordinates the one to the other. It is the "historical narrative" that he initially "had (...) the intention of not writing."

23　Stressing the omnipresence and the anthropological or cultural anchoring of narrative does not per se exclude a historically restricted approach to narrative. For example, Ansgar Nünning frequently urges scholars to historicize the narrative material they work with. In the case of travel writing, Nünning is especially interested in the historical prefiguration of the genre. See Nünning, Ansgar, "Zur mehrfachen Präfiguration / Prämediation der Wirklichkeitsdarstellung im Reisebericht: Grundzüge einer narratologischen Theorie, Typologie und Poetik der Reiseliteratur." In Points of Arrival: Travels in Time, Space, and Self, Zielpunkte: Unterwegs in Zeit, Raum und Selbst, edited by M. Gymnich et al., Tübingen: Francke, 2008, 11–32. Nünning, "Wie Erzählungen Kulturen erzeugen," 15–53.

24　Humboldt, Personal Narrative, 1, on p. xxxviii.

25　In the course of the travelogue, Humboldt actually uses two terms—historical and personal narrative—synonymously; sometimes even referring to this type of travel account as an itinerary. In this first inquiry into Humboldt's discourse on and practices of narrating, this paper follows the trail of the term historical narrative.

Earlier in the text, he even describes his decision to actually write a historical narrative as a pragmatic process of overcoming his "repugnance to write the narrative of my journey."[26] These reviews of Humboldt's initial opinion on travel writing certainly articulate a preference for descriptive writing. They also provide some insight into what he actually associates with these forms of writing. For him, a historical narrative unfolds the events of a voyage "in the order in which they successively presented themselves," meaning in the linear and thereby chronological order in which they succeeded each other. In this sense, he characterizes the historical narrative mainly through its temporal sequentiality and its focus on events. On the one hand, Humboldt's choice of criteria brings his definition close to contemporary narrative theory, which, in many cases, refers to both temporality and an event-like nature as characteristics of narrative.[27] On the other hand, the fact that Humboldt assigns a chronological order to the historical narrative indicates a more restricted definition of narrative—at least in the genre of travel writing.

Contrasted with narrative, description seems to be a relational way of ordering and presenting a voyage's results. It abandons the temporal order of the voyage, and focuses instead on the "relation" between the observed phenomena. This definition already hints at Humboldt's affinity for comparing that is the subject of the next section: It is comparison that he identifies in the text as one of the central means of uncovering universal laws and producing relational knowledge. Nonetheless, it should be stressed at this point that Humboldt depicts narrative and description as opposing ways of presenting a voyage's results. Whereas narrative organizes observations in a *chronological order*, comparative description presents the *relations* between different observations. The fact that narrative is not focused on universally valid relations—that it is, in other words, bound to particularities—gives Humboldt enough grounds to dismiss it as a mode of writing that is counterproductive to the epistemological goals of his travelogue.

At a later point in the travelogue, Humboldt gives a more detailed definition of the historical narrative that significantly expands the catalogue of criteria while, at the same time, blurring the lines between different modes of writing:

---

26   Humboldt, Personal Narrative, 1, on p. xxxix.
27   Martínez, "Was ist Erzählen," 2-5. Fludernik, Introduction to narratology, 6. Bal, Narratology, 79ff. and 214. Ryan, "Toward a definition of narrative," 28f.

An historical narrative embraces two very distinct objects; the greater or less important events that have a connection with the purpose of the traveler, and the observations which he has made during his journey. The unity of composition also, which distinguishes good works from those on an ill constructed plan, can be strictly observed only when the traveller describes what has passed under his own eye; and when his principal attention has been fixed less on scientific observations, than on the manners of a people, and the great phenomena of nature. Now, the most faithful picture of manners is that, which best displays the relations of men toward each other. The character of savage or civilized nature is portrayed either in the obstacles which a traveller meets with, or in the sensations which he feels. It is the man himself that we continually desire to see in contact with the objects that surround him; and his narration interests us the more, when a local tint is spread over the description of the country and its inhabitants.[28]

Again, Humboldt is identifying the chronological account of events as a specific quality of a historical narrative. It is the representation of "events" and "observations" that the traveler "made during the journey" that defines the narrative. But Humboldt adds two further criteria concerning the focal point of the narrative and the narrator's relationship to the world he narrates. Once more, Humboldt's criteria resemble criteria that contemporary narrative theory frequently discusses as crucial to narrativity—referred to mostly as *focalization* and *voice*.[29] Whereas the term *focalization* addresses the question whose perception or vision is represented in a narrative, the term *voice* refers to the question who is speaking in a narrative and whether or not that narrator is embedded in the world he narrates. Again, Humboldt's grasp of the narrative's focal point and the status of the narrator appears rather restricted: First of all, according to Humboldt's definition, a historical narrative should be internally focalized. He binds the account of events to the perspective of the traveler, to a focal point determined by "what has passed under his own eye" and "the sensations which he feels." But not only the focalization but also the voice of the narrative is character-bound and thereby limited. It is the "traveller" who narrates the story, who "describes" the events of the voyage. Because of that specification, the narrator can appear only as an embodied part of the world

---

28   Humboldt, Personal Narrative, 1, on p. xlf.

29   Fludernik, Introduction to narratology, 37ff. Bal, Narratology, 145ff. Martínez and Scheffel, Einführung Erzähltheorie, 66ff.

he narrates, as a "character-bound narrator"[30] or a "homodiegetic narrator."[31] This limits the scope of the narrator's voice: Strictly applied, he can recount only events he has been part of and observations he has made himself. According to Humboldt, linking the focalization and the voice of the narrative to the "traveler-narrator" grants "the unity of composition" and thereby the quality and authenticity of a historical narrative.

At first sight, these characteristics—temporal sequentiality and a determined, character-bound voice and focalization—seem to clearly distinguish the historical narrative from a description that is not limited by such narratological boundaries. But on closer inspection, Humboldt's definition partly challenges the strict distinction he establishes elsewhere in the text. Some of the terms that he usually contrasts with the historical narrative now reappear here in its very definition. For example, one of the most rewarding subjects of a historical narrative for Humboldt is the "relations of men toward each other." Whereas he previously seemed to look for relations in a descriptive mode, he now declares them to be one of the main goals of the historical narrative. Moreover, the historical narrative seems to contain "descriptions" in itself: for example, in the above-mentioned "descriptions of the country and its inhabitants." Humboldt's strict distinction between the two does not seem to hold up on closer inspection. In fact, just a few pages later he writes:

> In order to give greater variety to my work, I have often interrupted the historical narrative by simple descriptions. I first describe the phenomena in the order in which they appeared; and I afterward consider them in the whole of their individual relations.[32]

Whereas historical narrative and description appeared previously as mutually exclusive representational modes, they now appear as complementary modes of writing that can be incorporated into the same text. Following this logic, the term "historical narrative" refers to both the genre and a specific form of writing. The historical narrative, so to speak, contains both a historical narrative and descriptions. It shifts between linear storytelling from the traveler's perspective, on the one hand, and relational descriptions of individual phenomena, on the other.

30   Bal, Narratology, on p. 21.
31   Martínez and Scheffel, Einführung Erzähltheorie, on p. 84f.
32   Humboldt, Personal Narrative, 1, on p. xiii.

However, it is not only Humboldt's partial inconsistency concerning the differences between narrative and descriptive writing that raises doubt. His reflections can also be analyzed with regard to the prevailing forms of narrating and describing in the genre of scientific travel writing in the eighteenth and nineteenth centuries. Under the assumption that, to some extent, "genre directs the ways in which we *write, read,* and *interpret* texts" and even *"prescribes* artistic practices,"[33] Humboldt's repeated contrasting of historical narrative and comparative description must seem questionable. As Nünning argues, genre is one of the factors prefiguring the narrative and aesthetic make-up of travel writing.[34] Research on influential predecessors of Humboldt suggests that the constant shifting between narrating travel events and describing individual phenomena had become conventional by the beginning of the nineteenth century. For example, Brian W. Richardson argues that there is a strong routine and even a certain "boredom" of comparative description in James Cooks' travel writing.[35] Considering Cooks' influential role in the genre, being a travel writer who produced "a collection, a baseline, of texts on which all subsequent navigators can and must build,"[36] Humboldt's extensive commenting on the matter seems remarkable and anything but necessary. The fact that he frequently contrasts narrating and describing appears odd, given that the travel genre conventionally allows and even encourages the combination of the two.

Humboldt's separation of the two seems dubious on yet another level. From a contemporary narratological point of view, description does not appear as the counterpart of narrative, but rather as an integral part of it. According to Bal, description is "a particular textual form, indispensable, indeed, omnipresent in narrative," appearing as "a privileged site of focalization" that strongly influences the ideological and aesthetic make-up of a nar-

---

33   Pyrhönen, Heta, "Genre." In The Cambridge companion to narrative, edited by D. Herman, Cambridge: Cambridge University Press, 2007, on p. 109.

34   Nünning, "Präfiguration / Prämediation der Wirklichkeitsdarstellung im Reisebericht," 11–32.

35   Brian W. Richardson, Longitude and Empire: How Captain Cook's Voyages Changed the World, Vancouver: UBC Press, 2005, 147-157.

36   Richardson, Longitude and Empire, on p. 11.

rative.[37] Bal's evaluation of the role of description in narrative seems to apply to Humboldt's case: As the next section of this paper will show, Humboldt uses the descriptive passages of his travelogue to discuss the universal epistemological claims of the entire narrative. In this context, we have to ask as follows: if he emphasizes the role of description in narrative and blurs the criteria of narrative in the first place, why would he then insist on a strict distinction between narrating and describing? I argue that his differentiation of the two can be appropriately interpreted only when read in the context of *epistemological* issues. His preference for descriptive writing is not just a preference of genre: It corresponds with Humboldt's *epistemological* program. This connection becomes apparent when he defines the purpose of his travels as follows:

> [B]ut preferring the connection of facts which have been long observed, to the knowledge of insulated facts, although they were new, the discovery of an unknown genus seemed to me far less interesting than an observation on the geographical relations of the vegetable world, on the migration of the social plants, and the limit of the height which their different tribes attain on the flanks of the Cordilleras.[38]

Humboldt's preference for a relational text structure corresponds with his emphasis on relational knowledge—knowledge that focuses on the *connections* between observed phenomena and on the *universal laws* that these connections might disclose and not on the *chronological order* in which these phenomena have been observed by the traveler.[39] As the next section will clearly show,

---

37   Bal, Narratology, 35f. In a similar manner, Wolf Schmid argues that description and narration are not mutually exclusive text types but are rather organized as opposite ends of a scale. Narrative usually includes descriptive parts and vice versa. See Schmid, Elemente der Narratologie, 6f.

38   Humboldt, Personal Narrative, 1, on p. iv.

39   Ottmar Ette briefly draws attention to the role that relational logic and global comparison play in Humboldt's conception of science without inquiring further on the nature of Humboldt's comparative approach and its specific epistemological indications. See Ette, Ottmar, Alexander von Humboldt und die Globalisierung: Das Mobile des Wissens, Frankfurt a.M.: Insel, 2009, 15 and 23. On the universal scope and claim of Humboldt's epistemological project, see also Görbert, Johannes, Die Vertextung der Welt: Forschungsreisen als Literatur bei Georg Forster, Alexander von Humboldt und Adelbert von Chamisso, Berlin: De Gruyter, 2014. Hey'l, Bettina, Das Ganze der Natur und die Differenzierung des Wissens: Alexander von Humboldt als Schriftsteller, Berlin: De Gruyter, 2007.

Humboldt uses a discourse on genre, specifically on narrating, comparing, and describing, to discuss epistemological questions.

3.   Shifting from Narrating to Comparing

In this context, moments of transition in the text are of particular interest. Because Humboldt justifies at great length why he shifts from one form of writing to the other, these moments of transition might be interpreted as moments of crisis from a praxeological point of view. As Christian Bueger argues, knowledge that informs the performance of a practice usually stays invisible. Implicit knowledge becomes explicit only in moments of crisis when actors can no longer perform this practice without problems or when they start to perceive that practice as unsuitable to their needs.[40] Humboldt's rhetoric strongly suggests such a moment of crisis; or, more specifically, a disapproval of an established practice: namely, narrating things within travel writing. Much of the epistemological argumentation emerges from the central role that *comparison* plays in the descriptive passages.

For example, in his chapter on Tenerife, Humboldt dedicates many pages to the island's volcano, the Pico del Teide. After a linear account of his ascent to the volcano in the tradition of the historical narrative, Humboldt switches to a systematic geological description in which he compares the volcano to a great number of other volcanos all around the world. He introduces this transition as follows:

> Not to interrupt the narrative of the excursion to the top of the Peak, I have said nothing of the geological observations I made on the structure of this colossal mountain, and on the nature of the volcanic rocks of which it is composed. Before we quit the Archipelago of the Canaries, I shall delay a moment, and bring into one point of view what relates to the physical picture of these countries.[41]

Humboldt does not introduce the following comparative description as an integral part of the historical narrative. On the contrary, he depicts this mode

---

40   Bueger, Christian, "Pathways to practice: Praxiography and international politics," European Political Science Review 6(3), 2014: 395-397.

41   Humboldt, Personal Narrative, 1, on p. 196f.

of writing as a potential obstacle, as something that might "interrupt the narrative" and thereby interfere with the sequential recounting of events. Again, historical narrative and comparative description appear to be mutually exclusive modes of writing—mainly because the comparative description will "delay" the progress of the narrative and thereby disrupt its temporal structure. In contemporary narratological terms, the comparative description appears to *pause* the story time.[42] It does so by temporarily delaying the linear narration of travel events to focus on observations made in a specific scientific field of interest; in this case, the "geological observations" on the Pico del Teide.

At the end of this comparative text passage, Humboldt's distinction becomes slightly more complex. Here he integrates epistemological issues in his argument, commenting on the function that comparison has in the acquisition of knowledge:

> I have endeavoured to render these researches interesting, by comparing the phenomena of the volcano of Teneriffe with those that are observed in other regions, the soil of which is equally undermined by subterranean fires. This mode of viewing Nature in the universality of her relations is no doubt prejudicial to the rapidity suitable to an itinerary; but I thought, that, in a narrative, the principal end of which is the progress of physical knowledge, every other consideration ought to be subservient to those of instruction and utility.[43]

Humboldt elaborates on what exactly it is about the descriptive passages that interferes with the historical narrative. It is comparing on a global scale that interrupts the temporal structure of a narrative; or, more precisely, hinders the "rapidity suitable to an itinerary." Humboldt not only comments on what is "suitable" or useful to a certain genre, but also attaches a certain epistemic usefulness to different forms of travel writing. Comparing first leads to the acquisition of knowledge itself ("the progress of physical knowledge"); and second, to a perception of nature that takes into account "the universality of her relations." Narrating and comparing no longer appear to be mutually exclusive as different forms of travel writing but also as different forms with which to produce knowledge. Humboldt aims to "[view] the Globe as a whole" and to

---

42   Fludernik, Introduction to narratology, 32–34. Bridgeman, Terese, "Time and Space." In The Cambridge companion to narrative, edited by D. Herman, Cambridge: Cambridge University Press, 2007, 58.

43   Humboldt, Personal Narrative, 1, on p. 229f.

identify "laws" and "relations" between phenomena all around the world.[44] In another comparative passage of his travel account, he points out specifically that this kind of knowledge depends on a certain mode of travel writing:

> The form of a personal narrative, and the nature of its composition, are not well fitted for the full explanation of phenomena, which vary with the seasons, and the position of places. In order to study the laws of these phenomena, we must exhibit them in groups, and not separately, as they were successively observed.[45]

Here, Humboldt presents the *historical narrative* as a potential obstacle. Its sequential order appears to disrupt the search for connections and laws that link different phenomena. Narrative presents phenomena "separately" and not "in groups," meaning a focus on particularities and not on relations. In yet another comparative description, Humboldt again specifies that it is *comparing* that achieves his epistemological goal:

> Each part of the Globe is an object of particular study; and when we cannot hope to penetrate the causes of natural phenomena, we ought at least to endeavour to discover their laws, and distinguish, by comparison of numerous facts, what is constant and uniform from what is variable and accidental.[46]

In the logic of Humboldt's argument, comparing leads to the discovery of laws, links, and perhaps even causalities; and it does this on a global scale. Comparing brings about a synoptic view of all the world's phenomena—a synoptic view that guarantees the discovery of universal connections. Therefore, his preference for what he calls "systematic," "descriptive," or "comparative" travel writing is not just a genre preference but can be interpreted as an epistemological preference for relational, systematic knowledge that explains phenomena in the context of their global relations to each other. In contrast, narrative appears as an almost deficient mode of travel writing with a smaller epistemic scope that refers to the linearity of events and thereby only to singularities.

---

44   Humboldt, Personal Narrative, 1, 230.
45   Humboldt, Personal Narrative, 2, on p. 48.
46   Humboldt, Personal Narrative, 2, on p. 221.

4.   The Narrativity of Comparison: Volcanic Revolutions

If one looks at the actual descriptive parts of the text, Humboldt's comparative approach at first sight seems to fit with his programmatic distinction between historical narrative and comparative description. Comparing renders
the scope of the text more global. By comparing, Humboldt leaves behind
the spatial and temporal boundaries of the travel route and creates a global
geography that carries scientific meaning. For example, his comparative description of the Pico del Teide almost reads as an account of all the world's volcanoes. He compares the Pico del Teide to Mount Etna, Vesuvius, and Strom-
boli in Italy; to Cotopaxi and Tungurahua in Ecuador; and to Popocatépetl and
Pichincha in Mexico—to name just a few volcanoes he mentions for his global
comparisons.[47] Comparing enables him to address issues of general interest
and global consequence. One of the general issues in which he is interested is
the relationship between a volcano's size and the frequency of its eruptions.
He addresses this question as follows:

> The eruptions of the Peak [of Teide] have been very rare for two centuries
> past, and these long intervals appear to characterize volcanoes highly ele-
> vated. The smallest of the whole, Stromboli, is almost always burning. At
> Vesuvius, the eruptions are already rarer, though still more frequent than
> those of Etna and the Peak of Teneriffe. The colossal summits of the Andes,
> Cotopaxi, and Tungurahua, scarcely have an eruption once in a century. We
> might say, that in active volcanoes the frequency of the eruptions is in the
> inverse ratio of the height and the mass. The Peak [of Teide] also had seemed
> extinguished during ninety-two years, when, in 1798, it made its last erup-
> tion by a lateral opening formed in the mountain of Chahorra. In this interval
> Vesuvius had sixteen eruptions.[48]

These global comparisons create a synoptic view that takes into account not
only the particular volcano that is currently being visited in the narrative,
the Pico del Teide, but a great number of other volcanoes all over the world.
Humboldt is able to offer a causal hypothesis on the relationship between size
and eruptions, because he removes these volcanoes from their original spatial and temporal context and orders them according to their similarities and
differences. This comparative focus on possibly *global* geological causalities

---

47   Humboldt, Personal Narrative, 1, 196ff.
48   Humboldt, Personal Narrative, 1, on p. 248ff.

contrasts with the sequentiality of the travel narrative. It is also no longer restricted to the traveler's observations during the journey. Instead, Humboldt introduces additional information acquired at other times or by other travelers such as the number of eruptions of Italian volcanoes or the number of eruptions that occurred before or after the journey.

However, while the traditional sequential order of the historical narrative is interrupted, the text implies a sequential order on another level. Nature itself appears to be sequential. Humboldt refers to specific events, the "eruptions" of the named volcanoes, and he implies their sequential order by referring to their structure in time as "intervals." Volcanoes appear here as seemingly collective, nonanthropomorphic actors that are typical for factual forms of narrating.[49] In addition, the chain of volcanic events does not present itself as a random chronological succession. Humboldt attributes a pattern to what he reports: "We might say, that in active volcanoes the frequency of the eruptions is in the inverse ratio of the height and the mass." Thereby, Humboldt turns a random chronology into a causal sequence. Even though Humboldt usually thinks that the search for universal laws should be done by comparing and describing, here narrativity seems the obvious form of identifying a universal law: it links the reported events to a general cause. Thereby, Humboldt builds an alternative narrative that does not represent the *traveler's* experiences but *geological* events.

This temporal ordering of nature reappears in Humboldt's description of the Pico del Teide and its geological history. Humboldt is interested in geological changes over long periods of time that he describes as follows:

> Every thing indicates, that the physical changes, of which tradition has preserved the remembrance, exhibit but a feeble image of those gigantic catastrophes, which have given mountains their present form, changed the positions of the rocky strata, and buried seashells on the summit of the higher Alps. It was undoubtedly in those remote times, which preceded the existence of the human race, that the raised crust of the Globe produced those domes of trappean porphyry, those hills of isolated basalt on vast elevated plains, those solid nuclei which are clothed in the modern lavas of the Peak, of Etna, and of Cotapaxi. The volcanic revolutions have succeeded each other after long intervals, and at very different periods.[50]

---

49   Fludernik, "Narratologische Probleme des faktualen Erzählens," 120.
50   Humboldt, Personal Narrative, 1, on p. 240f.

Again, Humboldt adopts a comparative perspective by complementing the description of the Pico del Teide with comparative references to other volcanoes—in this case, Etna and Cotopaxi. Comparing enables to focus on global geological developments. This spatial expansion of the description's perspective is once more accompanied by a sequential ordering of nature. Again, Humboldt refers to specific natural events, to "volcanic evolutions" and "catastrophes," and he also arranges them again in a sequential order by stating that they "succeeded each other" and by referring to their order in time as "intervals" and "periods." The use of tense implies a double temporality, which is characteristic to narrative, a temporality on the level of the *histoire* as well as on the level of *discourse*. This double linearity is also stressed by Humboldt's reference to "remote times, which preceded the existence of the human race"—a reference that establishes a second timeline not only before "the human race" but also before the timeline of the story. Again, Humboldt replaces the historical narrative of the voyage with an alternative narrative of natural events. Comparing therefore breaks with the traditional setup of the historical narrative, but it involves narrative functions on another level—in this case, the sequencing of natural events. This temporal ordering of nature is not just a formal arrangement but serves as a way to create scientific meaning. By sequencing natural events, no matter how generally and vaguely, Humboldt identifies those events as part of a mutual causal process, the shaping of the earth and its "crust."[51]

It is important to note that the character-bound focalization that Humboldt finds necessary for the historical narrative also occurs in these descriptive text passages. Even though global comparisons broaden the scope of the text passage, Humboldt often authenticates these comparisons by referring to what he himself has seen, read, heard, or witnessed; for example:

---

51   As recent research on narrativity in the sciences has shown, scientific explanations become especially difficult when they are concerned with "historical events in nature"; and that, in many cases, "only narratives of the (special, singular, historical) events can be produced." Humboldt's efforts to establish a history of geological events suggest a similar epistemic situation and problem. However, whether or not we can actually assume a certain historical continuity here could be decided only by a further inquiry into narrating in the nineteenth and twentieth centuries. See Fuchs, Hans, "Typology of Uses of Narrative in Science: From Positioning Science in Culture through Creating Affect to Providing Explanations and Suggesting Concepts," narrativescience.org, www.narrativescience.org/Argument/Argument_Fuchs_01.html (accessed May 4, 2018).

> From the information of several well instructed persons, to whom I addressed myself, I found, that there are calcareous formations in the Great Canary, Fortaventura, and Lanzerota.[52]
>
> Of all the written testimonies, the oldest I have found of the activity of this volcano dates from the beginning of the sixteenth century.[53]

But there are shifts of focalization to be noted in the descriptive parts. Referring to events in "remote times" before humankind implies a less limited focalization—a perceptional scope that is not bound to any single agent in the universe of the narrative but involves times and places beyond the individual perception of a single character. However, when he reports those very events that are not *directly* accessible to the traveler-narrator as something he has *read* or *heard* from other sources, as shown above, Humboldt often uses a focalized perspective bound to the narrator. But there are also moments in the text when this limitation is momentarily replaced by an external focalization that is not bound to the perception of any character in the world of the narrative. For example, Humboldt writes:

> From those dark times, when the elements, subjected to the same laws, had not yet attained their present equilibrium, I come back to a period less tumultuous, nearer our own age, and on which tradition and history may throw some light.[54]

With a single small comment, "I come back," the narrator moves effortlessly from seemingly prehistoric events to recent history. This move through time and space can be interpreted as a shift of perception, a shift of gaze that cannot be ascribed to the character of the traveler. The narrator appears as a temporally and spatially flexible witness to natural events. This shift of focalization is linked to a shift of voice. For a moment, it is not the traveler who speaks but a heterodiegetic narrator who is located outside the world of the narrative. This narrator is able to speak about events beyond the traveler-narrator's knowledge, events that are of relevance to the general history of nature.

In conclusion, these examples show that comparing is closely intertwined with narrating. They also show that the specific narrative qualities of the comparative passages touch the epistemological scope of the text. A combination

---

52   Humboldt, Personal Narrative, 1, on p. 236.
53   Humboldt, Personal Narrative, 1, on p. 246.
54   Humboldt, Personal Narrative, 1, on p. 242.

of sequencing nature and shifting between different types of focalization allows a synoptic view of nature, its laws, and most importantly: its history. Humboldt puts a lot of energy in contrasting historical narrative and comparative description while articulating a strong distaste for narrative. He justifies this hierarchy by referring to his epistemological goal of uncovering global causalities and laws—a goal that does not match the particularity that is essential to narrative and its temporal sequentiality. Only when nature itself appears to be organized as a sequence of events does Humboldt's quest for universal natural laws become compatible with the particularity of narrativity. This sequencing of nature requires a shift in voice and focalization, a shift from a homodiegetic to a heterodiegetic narrator, as well as from a character-bound to an external focalization. In this way, narrativity, in spite of its particularity, can in fact strengthen a synoptic perspective and thereby contribute to Humboldt's search for universal or at least global laws and relations.

On the Narrative Order of Experimentation

Hans-Jörg Rheinberger

The claim to do nothing else than to let the things themselves tell their stories has a long tradition in the sciences. The venerable metaphor of the legibility of the world and of the letters in which the book of nature is written plainly corresponds with that demand of self-exposure. As Hans Blumenberg has shown, it has accompanied the sciences from the early modern times to the present, from the mathematical vision of Galileo Galilei to the letter-universe of the Human Genome Project.[1] The demarcation criterion for a discourse that can rightly claim to be scientific would thus be to allow things to express themselves according to their own grammar and their own lexicon. Succeeding in creating such a space of self-exposure would render scientific discourse transparent, and the congenial knowledge would be one that is essentially undistorted by the medium of its representation. To put it in another way: It would coincide with that representation. The question would thus not so much be whether scientific texts do narrate or not. Their scientificity would not consist in the fact that they would operate, in contrast to a descriptive narration, in the mode of an explanation, or according to different, but equivalent epistemological distinctions. Scientific texts would rather distinguish themselves from the many and multiple, invented or true stories that we tell ourselves about anything and everything, by the fact that they have *another author*. What I would like to do in this paper is to give this vision a particular twist: In trying to subvert it, I will take it up in a peculiar way.[2]

Posing the question of narration with respect to scientific knowledge thus means not only to pose the question of its content, or object, but in the last

---

1   Blumenberg, Hans, Die Lesbarkeit der Welt. Frankfurt am Main: Suhrkamp, 1981.
2   For an early and preliminary exposé of the following, see Rheinberger, Hans-Jörg, „Noch etwas über die experimentelle Ordnung der Dinge." In Wissenschaft und Welterzählung: Fakt & Fiktion. Die narrative Ordnung der Dinge, edited by M. Michel, Zürich: Chronos, 2013, 270-271.

instance, the question of the *subject* of the sciences. In trying to answer this question, one sooner or later faces the alternative that can be formulated as follows. In both there is narrative, albeit of a different author: Either one chooses the line sketched above and thus makes the scientist disappear behind the transcendence of the divine—or secular—order of things; then the subject of the sciences are the objects announcing themselves in their proper idiom. Or one opposes this kind of objectivism and aligns the scientific discourse about the order of things with the stream of all those stories that *we* tell us ourselves. As a consequence, according to temperament, the sciences present themselves as one grand or many small narrations. This is the classical dividing line.

In *Toward a History of Epistemic Things* I attempted to show that the *experimental* order of things is realized in a dynamic process condensed in experimental systems.[3] An attentive historical consideration of experimental systems opens a perspective pointing beyond the noted dichotomy. Experimental systems can be seen as the actual and actualizing technical setups for things of epistemic interest, for objects to become constituted as epistemic entities at all. A context of this kind is needed in order to endow an object with the very character of a thing of knowledge. For the experimenter, this context presents itself primarily as an instrumental one, as a necessary condition of manipulation. At a second look, however, one realizes that this context always consists already of what can be called "sedimented" knowledge, taking up an expression of Edmund Husserl.[4] Research technologies are materialized knowledge environments. One can consider them as collectively authorized vehicles, carriers that in their relation to an epistemic object constitute something like trans-subjective generators of events and of surprises. They are machines for producing what could be called epistemo-differences.[5] Sedimented knowledge brought into a research configuration thus has agency.

Now, we can reasonably argue that the essence of history is canalized contingency. History lives from and through events. Without boundary condi-

---

3   Rheinberger, Hans-Jörg, Toward a History of Epistemic Things. Synthesizing Proteins in the Test Tube. Stanford: Stanford University Press, 1997.

4   Husserl, Edmund, "The origin of geometry." In Edmund Husserl's "Origin of Geometry": An Introduction, edited by J. Derrida, translated by J. P. Leavey, Jr., New York: Harvester, 1978.

5   Wilhelm Johannsen talked about "genodifferences" as the epistemic objects of the genetic experimental systems of his time. Johannsen, Wilhelm, "The genotype conception of heredity." American Naturalist 45, 1911: 129-159, on p. 150.

tions, however, an incident cannot present itself as an event. And without events, there is no historical process. Even so, there is no good story without a certain amount of surprising turns in the framework of a plot. And in a really good story, it is the plot itself that provokes the turns. In his book *Grains et issues*, published in 1935, Tristan Tzara, the Romanian writer and one of the leading figures of French surrealism, characterized his view of a good literary story as follows. The passage is titled "The Experimental Dream": "Thus the story follows, and spreads across the frame of a logical development that reduces itself to an account of successive facts, but leaving an irrational and lyrical remnant open for discovery. This, in turn, overflows the vessel intended for it, and at times engulfs and floods the base, the foundation, the traditional scaffolding of the story. It is a lyrical superstructure whose elements are derived from the base structure and which, once it is realized, impacts back onto that structure from the heights of its new power. Occasionally, its force intensifies to such an extent that it undermines the meaning of the structure, corrupts it, abolishes it, annihilates it in its essence."[6] This is how Tzara sees, in the realm of literary production, a scaffold that allows, as he put it, to "bring forth new events not foreseen by the original plan."[7]

In the realm of science, we know of such structures precisely as experimental systems. They provide the space for knowledge provoking contingencies, for epistemic surprises that are more than the spurious sparks and accidents of a lucky intuition. They are arrangements that both produce history through their temporal order and stories through the permanent shift and displacement of meaning that characterizes them. The history and the stories that the sciences bring forth are written by experimental systems, which we have to address as difference machines. They do not have a once-for-all shape: Case studies are needed to set them in a proper light. They are to expose what could be called a poetology of research.

There is a line by Michael Polanyi — who started his career as a physical chemist and later became a philosopher of science — a stance that is telling in many respects in our context. Marjorie Grene quotes one of its versions in her book *The Knower and the Known*. It is a statement about what might be

---

6   Tzara, Tristan, Grains et issues, edited, introduced and annotated by H. Béhar, Paris: Garnier-Flammarion, 1981, 155-156.

7   Ibid., 155.

called the *research situation*[8] and reads as follows: "This capacity of a thing to reveal itself in unexpected ways in the future, I attribute to the fact that the thing observed is an aspect of reality, possessing a significance that is not exhausted by our conception of any single aspect of it. To trust that a thing we know is real is, in this sense, to feel that it has the independence and power for manifesting itself in yet un-thought of ways in the future."[9]

From the perspective of the researcher, we could say that what we are dealing with is an act of delegation. Setting up an experimental system revolving around an epistemic object and exploring some of the inexhaustible aspects of its thingness means to undercut the traditional subject-object relation in the sense of a face-to-face relation between an observer and something being observed. In an experiment, the act of observing is mediated by a technical arrangement of sorts that one brings into interaction with the epistemic object. According to Blumenberg, the action at a distance that this implies lies at the very basis of conceptualization *überhaupt*. Research, then, is second order conceptualization. It focuses on the process of conceptualization itself. An interaction of this sort has to be crafted in a way that the outcome — the traces that the interaction leaves behind — is not completely determined in advance. If it were, we would be dealing with a demonstration and not a research experiment. A research experiment lives from its aspect of "un-conceptuality," to use Blumenberg's notion for this peculiar tension.[10] It results from the effort to expand the realm of the conceptual which, in the very same movement, always also risks to reveal itself inappropriate.

Epistemic entities are thus things that by necessity leave something to be desired. They represent a knowledge-generating relation to the world: We can call it epistemicity. It is exploratory, driven by the desire to find, not to assert what is already given. When the great French experimental physiologist of the nineteenth century, Claude Bernard, once confided to his laboratory notebook

---

8   On this and the following, see also Rheinberger, Hans-Jörg, "On epistemic objects, and around." In WdW Review: Arts, Culture, and Journalism in Revolt, edited by D. Ayas and A. Kleinman, Rotterdam: Witte de With Publishers, 2017, 376-381.

9   Polanyi, Michael, Duke Lectures (1964), Microfilm, Berkeley: University of California Press, 1965, Library Photographic Service, 4th Lecture, 4–5. Quoted in Grene, Marjorie, The Knower and the Known, Washington DC: Center for Advanced Research in Phenomenology & University Press of America, 1984, 219.

10 Blumenberg, Hans, Theorie der Unbegrifflichkeit, Frankfurt am Main: Suhrkamp, 2007.

that "where one is no longer in the position to know, one must find,"[11] he expressed this situation in an exemplary and succinct manner.[12] Experimenters are specialists in creating situations in which such finding becomes possible. The movement of finding in science neither obeys the logic of mere chance nor that of pure necessity. It obeys a logic of its own, composed of elements of both, and in so doing, undercuts the stochastic rigor of the former *and* the deterministic rigor of the latter. It is a peculiar engagement with the material world that, on the one hand, requires intimacy with the matter at hand, and on the other, disentanglement, the capacity of *Verfremdung*. It has become common to address the event-provoking character of research under the label of serendipity, and it is probably not by chance that the term, which Robert Merton smuggled into the discourse on science, can be traced back to a fairytale from Persia.[13]

A reminiscence that dates back to the time when I was working on my case study on the history of protein biosynthesis research may illustrate the future-oriented power of finding, which at the same time acts as a recursive narrative force. Paul Zamecnik, whose laboratory at the Massachusetts General Hospital in Boston was the focus of this study, had been invited on several occasions, from the late 1950s to the middle of the 1980s, to lecture on the achievements of his lab. It is striking to see how he reported on the way the main steps of the research trajectory of his group changed over the distance of twenty years. In 1958, at a time his work was just beginning to enter the limelight of emerging molecular biology,[14] he presented a story that described the main findings strictly along his experimental trajectory. He used a vocabulary that remained largely operational and reflected the techniques his group was using to dissect the biosynthetic process under investigation: Amino acids were "incorporated" into protein, reflecting the fact that radioactivity added

---

11   Bernard, Claude, Cahier de notes 1850–1860, présenté et commenté par M. Grmek, Paris : Gallimard, 1965, 135.

12   It may be of linguistic interest here to note that the unit of research that usually leads to a publication is "the finding."

13   Merton, Robert K., and Elinor Barber, The Travels and Adventures of Serendipity: A Study in Sociological Semantics and the Sociology of Science, Princeton: Princeton University Press, 2006. As it were, the title of the book sees its subject matter itself as dominated by the principle of serendipity.

14   See, e.g., Zamecnik, Paul, "Historical and current aspects of the problem of protein synthesis." In The Harvey Lectures 1958-59, New York and London: Academic Press, 1960, 256-281.

in the form of radioactive amino acids was found to be associated with protein in the course of the experiment. "Microsomes" and later "ribonucleoprotein particles" were identified as the sites of protein synthesis, the former term referring to a cellular fraction that could be sedimented at high speed, the latter to the chemical constitution of a purified fraction. "Soluble RNA" was found to take up amino acids, referring to a ribonucleic acid fraction that remained soluble during high speed fractionation, and so on. The terminology carefully remained at the level of the technical set-up of the experimental system. In terms of theory, it was deliberately non-committal. In 1979, thus twenty years later,[15] the vocabulary hat completely changed, Now, the laboratory had engaged in "ribosome" studies from its beginning in the late 1940s, discovered "transfer RNA" around the middle of the 1950s, and with that, contributed a decisive link to understand the "language of the gene and that of the protein," that is, the process of "translation" as it was delineated at the end of the 1950s. Now, the story was told as one of molecular biology from the very outset, in a form that would have been unthinkable at the point where it started.

In the second part of this paper, I would like to explore the historiographical consequences of this deliberately epistemic view on the scientific research process. If experimental systems are to be seen as units of making scientific events happen, they can and should, of course, also become units of historiographical narration. The challenge of such case studies is to escape the illusion created by mapping the historiography directly onto the historical dynamics of the process. Georges Canguilhem has warned insistently against such a conflation. In his seminal paper on the object of the history of the sciences, he pleads for a clear distinction between objects of nature, objects of the sciences, and objects of the history of the sciences. In his address to the Canadian Society for the History and Philosophy of Science in 1966 in Montreal, he framed their respective differences with the following words: "The object in the history of the sciences has nothing to do with the object of a science. The scientific object, constituted by methodical discourse, is secondary, however not derived with regard to the natural, initial object that one might call pre-text, in playing with the sense of that word. The history of the sciences occupies itself with these secondary, non-natural, cultural objects, but it is not derived from them, as little as they are derived from the first. The object of the historical discourse is, in effect, the *historicity* of the scientific discourse,

---

15   Zamecnik, Paul, "Historical aspects of protein synthesis." Annals of the New York Academy of Sciences 325, 1979: 269-301.

inasmuch as this historicity represents the effectuation of a project that is internally normalized, but traversed by accidents, retarded or diverted by obstacles, interrupted by crises, that is, moments of judgment and of truth."[16] In his paper, Canguilhem polemicizes against a historical narrative that would nothing but emulate the scientific object by making use of the vocabulary of the sciences themselves. What is thus needed is a vocabulary that tries to capture the *historical* nature of scientific development, the dynamics of a process that, according to Canguilhem, is "normalized" *and* "interrupted by crises" at the same time. It thus creates the conditions of its own regulation *and* the conditions of critical transcendence without which that kind of historicity would not exist. For the perspective from experimental systems, this means that we need to think about the conceptual historical tools of a particular kind of micro-history. It is a micro-history that stands in contrast to other forms of traditional micro-history that have their place in the overall agenda of the history of the sciences, such as biographical or institutional narratives.

The focus of this kind of micro-history is on the materiality of the research process, with particular attention to the aleatoric moments that emerge from it and that have the power to orient it toward unforeseen directions. What we observe here is a peculiar kind of relationship between material continuity and conceptual reorientation. Take the example of the early genetic work of Carl Correns that extended over half a decade between 1894 and 1900.[17] Correns started to cross varieties of corn as well as peas with the explicit idea in mind to produce a clear instance of xenia—the appearance of characters of the pollinating variety on the seed and fruit of the mother plant—and then eventually elucidating its physiological background. Four years into the process, his goal was subverted by the observation of a roughly 3:1 ratio between the characters of the original varieties in the second generation of self-pollinating pea hybrids—instead of any unambiguous instance of xenia. The material continuity of his crossing regime together with the careful recording of the results allowed him to re-read the entire experimental process in terms of a re-discovery of Mendel's laws and to focus his attention on their corroboration in the final round of crossings. What we observe here is the material

---

16   Canguilhem, Georges, « L'objet de l'histoire des sciences ». In Etudes d'histoire et de philosophie des sciences, Paris : Vrin, 1968, 9-23, on p. 17, my emphasis.

17   Compare Rheinberger, Hans-Jörg, An Epistemology of the Concrete, Durham and London: Duke University Press, 2010, chapter 4.

continuity of an ongoing experimental process and, at the same time, a complete replacement of the epistemic object followed through the vagaries of that process.

The examples could be multiplied by widely different variants that cannot be understood without peculiar attention to the intricacies of the respective experimental processes. Therefore, such micro-histories generally require that the historian have a laboratory record at his or her disposal which will allow her or him to zoom into the experimental turning points. In the particular case of Correns, the laboratory protocols did not only have the function of memorizing the results of his experiments, they became part and parcel of the experimental process as it went on. For the historian, they are the surrogate for the experimental process, that is, the paper form of the experimental narrative itself that must serve him or her as a foil for her or his own efforts to come to terms, in Canguilhem's sense, with the historicity of the scientific object at hand.

The vocabulary of a historical epistemology living up to this challenge is not to be found ready-made in the annals of traditional epistemology. It requires an ongoing effort for those who continue to be interested in such a micro-history of the "mangles of practice," to put it in the words of Andrew Pickering.[18] For a deeper understanding of the scientific practices in the different corners and niches of the scientific universe, this approach remains indispensable. And there is one other thing that needs to be considered here. If we are to arrive at an understanding of how scientific knowledge and other cultural forms of knowledge and knowledge production hang together, we need to approach them from below and not from the bird's view of the theoretical products of selected sciences. Case studies of this kind are the privileged places of narration of a very particular, metonymic character: The very telling of such microstories only makes sense if they point beyond themselves. Their way of generalization has the form of 'standing for.' This is, however, in the nature of *microstoria* and distinguishes it from just local stories. Microstories pretend to tell a lesson. In that sense, we could even compare their role for history with the role experiments play in the sciences. Each and every experiment is a concrete, singular event. But it is only accepted as an experiment worth of consideration if it can be looked at as an instantiation of a more

---

18   Pickering, Andrew, The Mangle of Practice, Chicago: The University of Chicago Press, 1995.

general state of affairs. Otherwise one would not take it to be more than just fancy.

Different time frames are in need of different objects of historiographical narration. The story of an experimental system usually does not exceed the career span of a particular scientist or group of scientists. It amounts to what George Kubler considers as a project or productivity cycle, or indiction period, typically more than ten and less than twenty years.[19] If we would like to assess the dynamics of scientific development over a longer period of time, we need to think about entities other than experimental systems to guide our narratives. In the context of the history of the empirical sciences, one way of approaching the meso-range of the next order of magnitude, that is, the order in the range of a century instead of a decade, is to look at what I like to call experimental cultures.[20] Using experimental cultures as an object of narration, on the one hand, preserves the focus on practice inherent in the micro-approach. On the other hand, it allows for an understanding of how particular spaces of scientific activity are being formed that transcend a specific laboratory with its more or less unique experimental setup. These areas have taken different shapes in the history of the modern sciences. Until early into the twentieth century, they tended to condense into disciplines usually subjected to more or less stringent social codices. Just to give one example: A latecomer in the family of biological disciplines at the end of the nineteenth century was genetics. It rested on an experimental regime that was taken over from the realm of breeding plants and animals and adapted to the specification of those units that were thought to be responsible for the expression of certain organismic characters: the genes. Its two practical prerequisites were the selection of pure lines, on the one hand, and the capacity of such lines to be crossed with different ones, so that the behavior of the different characters could be followed in the progeny. A unique experimental culture resulted. It combined the living with mathematics: It made experimental use of living organisms, and the outcome of the experiments lent itself to the mathematical precision instrument of statistics. The gene as a new epistemic object took shape in this experimental context. Its characteristics remained formal, however. The

19   Kubler, George, The Shape of Time. Remarks on the History of Things, New Haven: Yale University Press, 1962, 101-102.

20   Compare Rheinberger, Hans-Jörg, "Cultures of Experimentation." In Cultures without Culturalism, edited by K. Chemla and E. Fox Keller, Durham and London: Duke University Press, 2017, 278-295.

experimental regime of classical genetics presented no handle to follow them down to the material, molecular level. On the other hand, a new instrument arose that soon would proliferate into other disciplinary specialties of the life sciences: the creation of model organisms. They were to accompany the life sciences as a living instrument over the entire twentieth century.

If we look at the twentieth century, however, we witness the emergence of spaces that are much less stringently codified and organized than disciplines, whose boundaries are less rigid, and whose constituency is more ephemeral than what we associate with the concept of discipline. It is therefore not by chance that historical epistemologists have, from the early twentieth century onward, pointed to this phenomenon and tried to find a conceptual framework to describe it. Gaston Bachelard, for instance, used terms such as "canton" or "district" in the image of different quarters of a city.[21] A generation later, Pierre Bourdieu introduced the notion of "field" to characterize relatively coherent areas of social and cultural activity, including scientific practice.[22] In doing so, he simultaneously aimed to make scientific practice comparable to other forms of practice.

A good example for such an experimental culture is in vitro experimentation. Test tube biology had its origins in what came to be called biochemistry at the beginning of the twentieth century. Its aim was to create artificial environments for partial biological reactions. It turned the "inner milieu" of Claude Bernard into an outer milieu.[23] It also had its place, however, in cell biology and in microbiology. And in the middle of the twentieth century, it formed an essential part of emergent molecular biology. We thus see clearly that we are dealing with a space that is more fine-grained than that of discipline, and above all, one that has a different focus. With his notion of "cultures of emergence," Bachelard has pointed to the core of all such different, sub-disciplinary knowledge spaces: the eventuation of novelty.[24] Shifting one's narrative attention toward the specificities of these experimental environments and their potential of innovation means to keep practice in the center and, at the same time, finding a way to de-localize the story that is to be told. It

---

21   Bachelard, Gaston, Le rationalisme appliqué, Paris : Presses Universitaires de France, 1949.
22   Bourdieu, Pierre, Pascalian Meditations, Palo Alto : Stanford University Press, 2000, esp. Chapter 3.
23   Bernard, Claude, Leçons sur les phénomènes communs aux animaux et aux végétaux (1878-1879), Paris : Vrin, 1966, in particular the Second Lecture, section III.
24   Bachelard, Le rationalisme appliqué, 133.

entails another form of generalization than the one we encountered with experimental systems as micrological units of narration. Whereas their modus was metonymic, here the modus is analogical and parallactic.

Finally, what about macro-histories? Underneath experimental systems as well as cultures of experimentation, we witness a flux of time that obeys yet another order of duration, sometimes extending over several centuries. The question is what kinds of entities would be apt to serve as narrative guides over such extended periods of time? From the perspective of scientific practice, Foucauldian "discourse" comes to mind as a possible candidate to do the job. Following Michel Foucault, a discourse consists of an epochal, overarching set of practices and standards as well as the beliefs embodied in them that delimit what is conceivable and enunciable within that framework.[25] However, when the focus is on the dynamics of a particular realm of science in the making, one needs to be more specific. The sciences unfold in the context of discourses, but they are only spots of condensation within them. Thomas Kuhn has talked about a "disciplinary matrix" in this respect,[26] a structure into which paradigms are wired. But in either case, the focus is on closure, on what is excluded. What we are to be looking for here, however, is a narrative guide that would focus on the *openings* and displacements along a longer-term trajectory and that would not exclude the unprecedented at this temporal macro-level.

Again, Canguilhem can be helpful here. He suggested that such long-term histories might best be written as histories that follow the trajectory of scientific *concepts* from one realm of inquiry to another and to observe their vagaries and varying embodiments and instantiations.[27] The focus then lies on broader figures of change. Let us take the example of the concept of heredity and sketch its trajectory in extremely broad strokes.[28] A concept of heredity was absent from the space of natural history until the late eighteenth century. Theories of generation, be they preformationist or epigenetic, were not in need of such a concept. Around 1800, it entered the realm of the biological

---

25   Foucault, Michel, The Discourse on Language. Appendix to The Archeology of Knowledge, New York: Pantheon, 1972, 215-237.
26   Kuhn, Thomas, "Second thoughts on paradigms." In The Essential Tension, Chicago: The University of Chicago Press, 1977, 293-319.
27   Compare Canguilhem, Georges, La Formation du concept de réflexe aux XVIIe et XVIIIe siècles, Paris : Presses Universitaires de France, 1955.
28   Müller-Wille, Staffan, and Hans-Jörg Rheinberger, A Cultural History of Heredity, Chicago: The University of Chicago Press, 2012.

from the realm of the legal where it had its original place: the idea of material properties transmitted from one generation to the next. Concomitantly, the notion of generation completely changed its meaning. Conceptualizing generational change in terms of such properties—of which organisms acted as their carriers—took shape over the course of the nineteenth century in different practical and discursive areas such as medicine, agricultural breeding, anthropology, and evolution. To begin with, it created a scattered epistemic space, one that only became unified toward the end of the nineteenth century. Along with this condensation, the epistemic space of heredity became compacted as an epistemic object sui generis: the gene. The gene then permeated all of twentieth century life sciences, and it inhabited a plethora of experimental systems and cultures of experimentation, thereby creating a number of successive auras around itself. They can be addressed as so many "images of knowledge."[29] The first of these images was that of an 'atom of life.' But as such an atom, it remained elusive. The second was that of a material 'information carrier.' Now the atom had materialized as a molecule, but what it meant to carry information remained elusive. The third image was that of an element of a 'map.' Now it became the node of a network, but the nature of that network remained to be determined. And the story goes on.

I have restricted myself to this example to characterize the transgressive power of the epistemic object 'gene' in the space of research—the privileged space of my concern in this paper—and I have neglected the onto-theological and technological stories that accreted around heredity with all their social, political, cultural, and medical consequences—those second-order life-world materializations of powerful images of knowledge from which the sciences can, of course, not be detached. Narratives at this level would talk, for instance, about the mechanization of the worldview, the geneticization of society or, in general terms, the scientification of our world picture. But with that, we would have reached a level of generalization that borders on what is being called "grand narratives." A generation earlier, its label was "ideology." This is a level that does not make sense from a perspective that focuses on the process of research. It is even counter-indicated, as long as one is convinced—of which I remain—that scientific exploration, together with a few other cultural activities such as art, is endowed with an ongoing and irresistible subversive power. To be subversive means nothing else than to have the power of resisting totalization.

---

29   Elkana, Yehuda, Anthropologie der Vernunft, Frankfurt am Main: Suhrkamp, 1986.

Now, a narrative is a narrative only as long as one can imagine that it might have been otherwise. Narration therefore comes with an intrinsic quantum of potential plurality, and therefore with an unavoidable amount of concreteness and circumstantiality. An abstract story—no less than an abstract experiment—is a contradiction in terms. This elective affinity is the ground on which experiment as narration and narration as experiment can come together and on which their paths can cross and inform each other.

The Flower People of Shanidar
Telling a New Tale about our Neanderthal Brothers

Oliver Hochadel

Neanderthals were artists! In February 2018 a publication in *Science* made headlines around the world. Using new dating techniques researchers claimed that simple paintings in three Spanish caves were much older than previously thought, namely over 60,000 years ago. And since *Homo sapiens* only settled much later in the Iberian peninsula they must be the work of Neanderthals.[1] Although met with some skepticism this recent finding may be understood as yet another step in the "upgrading" of the Neanderthal that began in the 1950s. Since mid-century the popular image of the Neanderthal as brutish and essentially subhuman started to give way to a vision of a much more sympathetic, intelligent and cultured creature. Each revision brought "them" closer to "us".

A crucial step in this promotion of the Neanderthal from beast to brother were the finds in the Shanidar cave in Northern Iraq made between 1951 and 1960. These discoveries reached a larger public through Ralph Solecki's book *Shanidar. The First Flower People* published in 1971. An unusually high amount of pollen found near a skeleton seemed to indicate the burial of a Neanderthal on a bed of colorful flowers. This hypothesis had a strong impact on popular culture. Novels, popular science books, dioramas in museums and a Hollywood movie exploited the powerful imagery of the deeply moving flower burial.

This paper will analyze the Shanidar case applying the double perspective of this volume. Firstly, narrating: The narratives that in one way or another originated from Solecki's book soon developed their own dynamic. What topics, stories and imageries were spun out of the Shanidar Neanderthals and to whom were they addressed? Secondly, comparing: Narratives by *Homo sapiens*

---

1   Hoffmann, D. L. et al., "U-Th Dating of Carbonate Crusts Reveals Neandertal Origin of Iberian Cave Art," Science 359 (6378), 2018: 912–915.

about Neanderthals are always, sometimes more, sometimes less explicitly comparative. The contrasting with "us" has been a constant feature of Neanderthal research ever since the first fossils emerged in 1856. How does the creature with the thick brow ridges, sloping forehead and supposedly stooped, shuffling gait live up to his supposedly more evolved sapiens brother?

1. The Neanderthal—from beast to brother

The changing image of the Neanderthal, alternating between ancestor, brother and distant cousin has been amply documented and commented on by both historians of science and practicing paleoanthropologists.[2] From the fossil discovery in the Neander Valley in 1856 it took about half a century for the acceptance of Neanderthal as a distinct species to *Homo sapiens*. Shortly afterwards, just before World War I, French prehistorian Marcellin Boule "expelled" the Neanderthal from the human lineage. His analysis of the La Chapelle-aux-Saints skeleton created the highly influential notion of a stooped figure. In popular representations the Neanderthal was also depicted with thick hair or even fur, and, if male, armed with a club. In the common imaginary he was brutish and certainly not particularly bright.

It took until the 1950s to revise Boule's erroneous judgment. The Neanderthal walked fully upright and he resembled us in many respects. As Straus and Cave put it in 1957: "if he could be reincarnated and placed in a New York subway-provided that he were bathed, shaved, and dressed in modern clothing-it is doubtful whether he would attract any more attention than some of its other ‚denizens'."[3] This thought-experiment, formulated in an academic article, proved to be very "catchy" and reverberated strongly in popular culture, generating cartoon images of Neanderthals wearing suits and sporting hats.

---

2   For a concise summary see Sommer, Marianne, "The Neandertals." In Icons of Evolution: An Encyclopedia of People, Evidence, and Controversies, edited by B. Regal, Westport: Greenwood, 2008, 139-166. For accounts by paleoanthropologists see Trinkaus, Erik, and Pat Shipman, The Neandertals. Changing the Image of Mankind, London: Jonathan Cape, 1993. Schrenk, Friedemann, and Stephanie Müller, Die Neandertaler, München: C.H. Beck, 2005.

3   Straus, William L., and A. J. E. Cave, "Pathology and the Posture of Neanderthal Man," The Quarterly Review of Biology 32, 1957: 348-363, 359.

Scholars have repeatedly pointed out that from the earliest publications and pictorial representations the Neanderthal not only led an academic life but also a public one. The Neanderthal has been omnipresent in popular culture for well over one hundred years now, in popular science books, newspapers, magazines and museum exhibits, but also in comics, novels, movies, all sorts of gadgets and as a metaphor usually meant as an insult, signifying someone who is hopelessly backward and out of touch.

In line with the "communicative turn" in history of science[4], a neat separation of "strictly scientific" and "merely popular" representations of the Neanderthal seems impossible. These two spheres were and are intertwined, feeding off each other. In fact, the movement "from beast to brother" became itself a narrative, that comprises this change in a nutshell. This narrative of close resemblance can be found in media ranging from newspaper articles to large exhibitions such as the recent "Néandertal: lui et nous" at the Musée de l'Homme in Paris in 2018.[5] To be sure, this movement was not linear. The 1980s saw some setbacks in humanizing the Neanderthal fully. Archaeologists such as Lewis Binford wondered whether they might not have been completely unlike us, lacking the capacity to plan ahead.[6] Erik Trinkaus emphasized their fundamental difference to us.[7] Others doubted whether Neanderthals were able to speak at all. After it had been confirmed that they actually were able to speak it was suspected though that they lacked complex language and symbolic capacities. More recently the difference between "him/her" and us has narrowed again. A perforated piece of bone found in Slovenia led to the (in the meantime much questioned) suggestion that Neanderthals played music on a flute. The remnants of putative Neanderthal art in Spanish caves mentioned at the beginning of this article has been the most recent step in this process of humanizing. Yet clearly the most impacting find was the analysis of the Neanderthal genome. In 2010, Svante Pääbo and colleagues showed that

---

4   Secord, James A., "Knowledge in Transit," Isis 95 (4), 2004: 654-672.

5   Catalogue of the exhibition: Depaepe, Pascal, and Marylène Patou-Mathis, editors, Néandertal: lui et nous, Paris: Gallimard, 2018.

6   Binford, Lewis R., "Human Ancestors: Changing Views of Their Behavior," Journal of Anthropological Archaeology 4, 1985: 292-327. Binford, Lewis R., "Isolating the Transition to Cultural Adaptations: An Organizational Approach." In The Emergence of Modern Humans, edited by E. Trinkaus, Cambridge: SAR/Cambridge University Press, 1989, 18-41.

7   Trinkaus and Shipman, The Neandertals, 417.

non-African humans have between 1,5 and 4 percent Neanderthal DNA.[8] Most of us carry them in us so to speak. This discovery also answered the much-debated question whether Neanderthals and *sapiens* had common offspring. The issue of "interbreeding" had been pursued and played out in a host of novels set in prehistory already way back, since the early twentieth century. The literary realm of paleofiction had become in a sense the "testing ground" for this research question imagining different kind of scenarios.[9]

As we shall see in the remainder of this article, the interaction between archaeological research and popular culture was particularly fruitful in the case of the Shanidar Neanderthals, creating a whole cosmos of moving narratives and powerful images.

2.   A flower burial and an elderly invalid

The source for this interaction was *Shanidar. The First Flower People*, published by the archaeologist Ralph Solecki (1917-2019) of Columbia University in 1971. For the most part the book is a description of four excavation seasons in Northern Iraq (1951, 1952/53, 1957, 1959/1960). In the cave of Shanidar in the Zagros mountains in Kurdistan, about 470 kilometers north of Bagdad, and close to the borders of Turkey and Iran, the remains of nine Neanderthals were found, considered to be around 60,000 years old. The cave itself is enormous, the opening is 25 meters broad and 8 meters high and extends up to 50 meters back into the mountain.

Apart from anatomical questions with respect to the variation of Neanderthal anatomy the discoveries allowed for inferences on the behavior of that species. Two of the skeletons unearthed stand out in this respect: Shanidar I and Shanidar IV. In her analysis of soil samples from the area where Shanidar IV was found, French paleobotanist Arlette Leroi-Gourhan (1913-2005) could identify a large number of pollen grains that stemmed from at least 6 different plants. This became Solecki's central argument for the humanity of Neanderthals and in a sense the main selling point of the book: "It is that early

---

8   Green, Richard E. et al., "A Draft Sequence and Preliminary Analysis of the Neandertal Genome," Science 328 (5979), 2010: 710-722.

9   De Paolo, Charles, "Wells, Golding and Auel: Representing the Neanderthal," Science Fiction Studies 27, 2000: 418-438. Ruddick, Nicholas, The Fire in the Stone. Prehistoric Fiction from Charles Darwin to Jean M. Auel, Middletown, Conn.: Wesleyan University Press, 2009.

Stone Age Man had very human feelings; to a very much greater extent than we have ever suspected. He was buried with flowers."[10]

Several of the individuals found were not buried, but rather had been presumably killed by rock fall from the roof of the cave caused by the frequent earthquakes in the region. Solecki includes a second case that bolsters his case for the humanity of the Neanderthals, the male individual known as Shanidar I—arguably the most famous invalid in prehistory. The number of lesions this human being had to endure—and survived—in his life-time is breathtaking, if not heartbreaking. His right arm was withered and mutilated (or even got partially amputated). At an early age he must have received a strong blow against the left orbit of his skull which probably left him blind on his left eye. Deformations in both his feet presumably prevented him from participating in hunting. Injuries and infections sustained during adolescence but possibly also in part a birth defect might have been the reasons for these severe handicaps. The fossils show that his lesions had healed and he continued to live with his disabilities to the age of 40 or more, considered an advanced age for a Neanderthal. As Solecki suggests, this crippled individual needed prolonged assistance from his group: "Although he was born into a savage and brutal environment, Shanidar I man provides proof that his people were not lacking in compassion."[11]

In the analysis of the Shanidar fossils Solecki relied on renowned anthropologist and paleopathologist T. Dale Stewart (1901-1997) and later on on human paleontologist Erik Trinkaus (*1942) who confirmed and expanded on Stewart's hypotheses.[12] With these two major discoveries, Shanidar I and Shanidar IV, Solecki felt confident to claim that Boule must have been wrong and that Neanderthals were in fact our ancestors.[13] That Neanderthals were fully human is the main message of his book—but not the only one.

10   Solecki, Ralph S., Shanidar. The First Flower People, New York: Knopf, 1971, 3.
11   Solecki, Shanidar, 195, also see 192–193.
12   Trinkaus, Erik and M. R. Zimmerman, "Trauma among the Shanidar Neandertals," American Journal of Physical Anthropology 57, 1982: 61–76. In 1983 Trinkaus published the most comprehensive examination of the Shanidar fossils to date: Trinkaus, Erik, The Shanidar Neanderthals, New York: Academic Press, 1983.
13   Solecki, Shanidar, 11, 270.

3.   The Iraqi Kurds, a fierce mountain folk

Solecki clearly thought of himself not only as an archaeologist but also as an anthropologist. The excavation team had recruited quite a number of men from the local population to do the digging. So Solecki pursued a kind of double field work, excavating Neanderthals but also closely observing the Kurds and their customs. The bibliography of his book bears testimony that Solecki read widely on the history of the region and its people. Already in 1955, well before the publication of *Shanidar. The First Flower People*, Solecki even published a record entitled "Kurdish Folk Songs and Dances".[14]

In his book, Solecki takes a genuine interest in the Kurds of Northern Iraq. Yet this does not save him from "orientalizing" them. He describes a people rich in culture but somehow "stuck" in time. "The history of the Kurds seems to be one of constant ferment without any real advance or progress."[15] For Solecki the Iraqi Kurds were "a simple and fierce mountain folk" ready to take up arms if their leaders told them to.[16] The Zagros mountains represent "a cultural backwater of the higher civilizations to the south"—both in the Paleolithic as well as nowadays.[17] Solecki repeatedly alludes to continuities between the deep past and the present. During the excavations in the 1950s, there were still local people living in the cave during the winter period seeking shelter from the elements—just like the Neanderthals and later on prehistoric *Homo sapiens* had done.[18] Solecki writes: "… the Shirwanis blossomed out with a kind of slipper or moccasin made of the skin of an animal, hair inside, which was tied around their ankles by string, in a style which harked back to prehistoric days."[19]

This idea of historic continuity is also present in the stratigraphy of the cave itself. There are at least 14 meters of sediments representing up to 100,000 years of settlement (including some discontinuities) with the

---

14   Kurdish Folk Songs and Dance, recorded by Ralph S. Solecki: Folkways Records 1955; also see Solecki, Shanidar, 146, 153.

15   Solecki, Shanidar, 80.

16   Solecki, Shanidar, 142, see already 82.

17   Solecki, Shanidar, 163; also see 173, 268.

18   Later on Solecki published an article about them: Solecki, Ralph S., "Contemporary Kurdish Winter-Time Inhabitants of Shanidar Cave, Iraq," World Archaeology 10, 1979: 318–330.

19   Solecki, Shanidar, 110.

Neanderthals being at the very bottom (level D) and later populations of *Homo sapiens* (level C to A) higher up in the deposit.

Of particular significance, given the title of the book, is the love for flowers that the Kurds harbor, as Solecki mentions on several occasions.[20] And once he draws a direct parallel between the past and the present: "Among such flower lovers are the modern Kurds, who are simply following a tradition of the country dating from about 60,000 years ago."[21]

4.   The impact of Solecki's book

Despite its catchy title *Shanidar. The First Flower People* was apparently not a commercial success. There was only one edition in the US followed by a British edition that appeared in 1972 under a different title.[22] Only a handful of reviews appeared[23] and only one translation (into Japanese).[24] Nevertheless the book proved to be influential, yet not in a straight-forward way. The title played an important role in this. As *Shanidar. The First Flower People* was published in 1971 its title clearly alluded to the hippies' "Flower Power" of the late 1960s. It seems reasonable to suggest that the title was chosen by the publisher to maximize sales. Maybe it was just an inevitable choice, trying to relate the main thesis of the book with the present.

Yet the title very much conditioned the way it was understood. Most later comments on Solecki's book point to the counter-culture of the 1960s as the essential historical background. Ruddick for example wrote: "Our hirsute cousins, it seemed, had been gentle, peace-loving creatures. In an era when shaggy-locked antiwar protesters waving flowers confronted helmeted, heavily armed militia, it seemed easy to imagine how the Neanderthals' extinction might have come about." And he claimed that this "rehumanization" of the

---

20   Solecki, Shanidar, 122, 156, 176.

21   Solecki, Shanidar, 269.

22   Solecki, Ralph S., Shanidar: The Humanity of Neanderthal Man, London: Allen Lane/The Penguin Press, 1972. The pagination of the British edition (fewer pages, more densely printed) is different, but its content is practically identical.

23   Morrison, Philip, Scientific American 225, 1971: 234–238. John Norris, The Science Teacher 39, 1972: 64. T. Cuyler Young Jr., The American Historical Review 80, 1975: 375–376.

24   Solecki, Ralph S., Shanidar Dōkutu No Nazo, trans. Yukinari Kōhara and Noriko Matsui, Tokyo: Sōju Shobō, 1977.

Neanderthals was "increasingly attuned to the countercultural currents in the United States during the Vietnam War".[25]

Such quotes indicate that *Shanidar. The First Flower People* was a book much cited but actually little read. Except for some of the early reviews nobody commented on the significant anthropological content of the book and the suggested continuity between Neanderthals and present-day Kurds. At the same time Solecki makes no allusion whatsoever to the flower people in California or the massive student protests at Columbia University in New York in 1968, where he taught at the time. The book does have a political background but it is an entirely different one: the turmoil of the Near East in the 1950s. Solecki mentions the Suez Crisis of 1956/57 and the end of the Iraq monarchy in 1958.[26] He could not continue with his research in the 1960s because of the armed conflict between government forces and the Iraqi Kurds.[27] In fact only a small portion of the sediments within the cave had been excavated and only very recently archaeologists have returned to find more fossils.[28]

Shanidar. The First Flower People was clearly meant to be a popular science book aimed at a general audience. As Solecki himself states in the foreword: "The human-interest details concerning the findings cannot be properly told in a severely technical paper."[29] He details the day-to-day of the excavation under precarious conditions and includes numerous anecdotes about the interactions of the US-American archaeologists with the local population. The reader learns about near-accidents, including the dramatic event of an earthquake during the dig. Solecki uses familiar literary techniques of the adventure-discovery-story: first person narrative, "local context" that adds an exotic flavor, and the thrill of the unexpected fossil find that supposedly changes fundamentally the understanding of our past. I personally found the book well written and intriguing. Nevertheless, *Shanidar. The First Flower People* did not

---

25   Ruddick, The Fire in the Stone, 71. Similar allusions in Shreeve, James, The Neandertal Enigma. Solving the Mystery of Modern Human Origins, New York: William Morrow, 1995, 53–54. Schrenk and Müller, Die Neandertaler, 60.

26   Solecki, Shanidar, 173.

27   Solecki, Shanidar, 3, also see Leroi-Gourhan, Arlette, "Shanidar et ses fleurs," Paléorient 24, 1998: 79–88.

28   Pomeroy, Emma et al., "Newly Discovered Neanderthal Remains from Shanidar Cave, Iraqi Kurdistan, and their Attribution to Shanidar 5," Journal of Human Evolution 111, 2017: 102–118; Pomeroy, Emma et al., "New Neanderthal Remains Associated with the 'Flower Burial' at Shanidar Cave," Antiquity 94/ 373, 2020: 11–26.

29   Solecki, Shanidar, xi.

match for example the bestsellers in the field of HOR that would follow a couple of years later. In 1977, Richard Leakey published *Origins* and in 1981 Donald Johanson followed suit with *Lucy. The Beginnings of Humankind*.[30] In this book, Johanson used a very similar formula to Solecki's, but in his case he landed an international bestseller turning the *Australopithecus afarensis* Lucy into an icon of HOR.

There is now a wide consensus among historians of science and STS scholars that science popularization entails much more than communicating recent findings to a larger public in a palpable way. In their popular works scientists often address their own peers circumventing the formal restrictions of peer-reviewed articles.[31] The format of the popular science book, offering ample space, is particularly apt for expounding new ideas, including controversial hypotheses.[32] The field of HOR lends itself in particular to expand the academic battle field due to the enormous and persistent interest of the general public in "their" origins.[33]

In *Shanidar. The First Flower People* Solecki forcefully proposes his thesis about the flower burial and its significance for revising our view of Neanderthals granting these "early men the full range of human feelings and experience".[34] Creatures who cared so much about their dead could only be human beings in the full sense of the word. Solecki's book was published in 1971, that is *before* some of the major scientific articles on the Shanidar Neanderthals.[35]

30   Leakey, Richard E., and Roger Lewin, Origins. What new Discoveries reveal about the Emergence of our Species and its possible Future, New York: E.P. Dutton, 1977. Johanson, Donald C., and Maitland A. Edey, Lucy. The Beginnings of Humankind, New York: Simon and Schuster, 1981.

31   Bucchi, Massimiano, "When Scientists Turn to the Public. Alternative Routes in Science Communication," Public Understanding of Science 5 (4), 1996: 375–394.

32   Sepkoski, David, "Paleontology at the 'High Table'? Popularization and Disciplinary Status in Recent Paleontology," Studies in History and Philosophy of Science Part C: Studies in History and Philosophy of Biological and Biomedical Sciences 45, 2014: 133–138.

33   Hochadel, Oliver, "Die Knochenjäger. Paläoanthropologen als Sachbuchautoren." In Sachbuch und populäres Wissen im 20. Jahrhundert, edited by A. Hahnemann and D. Oels, Frankfurt a.M.: Peter Lang, 2008, 29–38; Hochadel, Oliver, "A Boom of Bones and Books. The 'Popularization Industry' of Atapuerca and Human-Origins Research in Contemporary Spain," Public Understanding of Science 22 (5), 2013: 530–537.

34   Solecki, Shanidar, 250.

35   Leroi-Gourhan, Arlette, "The Flowers Found with Shanidar IV, a Neanderthal Burial in Iraq," Science 190 (4214), 1975: 562–564. Solecki, Ralph S., "Shanidar IV—a Neanderthal Flower Burial in Northern Iraq," Science 190 (4217), 1975: 880–881. Solecki, Ralph S.,

Although the excavations had come to a forced halt in 1960, the scholarly publications took a long time to come out.[36] Solecki and his colleagues published a couple of articles in the 1960s but the flower burial did not yet figure in these.[37] Then, in 1968 palynologist Arlette Leroi-Gourhan published an article suggesting that colorful flowers might have been put on top of the dead body of Shanidar IV.[38] In Solecki's words: Leroi-Gourhan "provided the spark for writing this book".[39] But the fully-fledged theory of a flower burial and the implications for our understanding of the Neanderthals were only set forth in Solecki's book. *Shanidar. The First Flower People* provides thus another example of how the medium of the popular science book serves to communicate new findings and potentially controversial interpretations to both, the general and the specialist public. Solecki's book was seriously read by scholars because they quote directly from it and take issue with some of his claims therein.[40]

In 1975, a more detailed analysis of the pollen by Leroi-Gourhan prompted Solecki to go even further, this time not in a popular science book but in a peer-reviewed article in *Science*. It seemed that the Shanidar Neanderthals attributed medicinal qualities to some of the flowers detected in the burial. "One may speculate that the (individual) was not only a very important man, a leader, but may have been a kind of medicine man or shaman in his group."[41] As we shall see below, this interpretation increased the potential of the flower

"The Implications of the Shanidar Cave Neanderthal Flower Burial," Annals of the New York Academy of Sciences 293, 1977: 114–124. Stewart, T. D., "The Neanderthal Skeletal Remains from Shanidar Cave, Iraq: A Summary of Findings to Date," Proceedings of the American Philosophical Society 121 (2), 1977: 121–165.

36   Or were published in the Iraqi journal Sumer and thus not easily accessible to most scientists, Stewart, "The Neanderthal Skeletal Remains," 121.

37   Solecki, Ralph S., "Prehistory in Shanidar Valley, Northern Iraq", Science 139 (3551), 1963: 179–193, is an overview article and does not discuss the Neanderthal finds in any detail yet.

38   Leroi-Gourhan, Arlette, "Le Néanderthalien IV de Shanidar," Comptes Rendus de la Société Préhistorique 65, 1968 : 79–83.

39   Solecki, Shanidar, xii.

40   Gargett, Robert H., "Grave Shortcomings: The Evidence for Neandertal Burial [and Comments and Reply]," Current Anthropology 30 (2), 1989: 157-190, 176; Pomeroy et al., "Newly Discovered Neanderthal", and Pomeroy et al., "New Neanderthal Remains".

41   Solecki, "Shanidar IV," 881. This article refers to Leroi-Gourhan, "The Flowers Found," published three weeks earlier in the same journal, Science.

burial for future adaptations in popular culture. All kinds of stories in different media could be spun out of the hypothesis of Solecki and Leroi-Gourhan.

After the publication of Solecki's book both the stories of Shanidar I and Shanidar IV in its basic form became staple items in popular science books on human origins or more specifically on the Neanderthals for the remainder of the twentieth and well into the twenty-first century.[42]

George Constable's short but amply illustrated *The Neanderthals*, published in 1973, was far more influential in spreading the ideas of Solecki and thus the "new" image of the Neanderthal than the book *Shanidar. The First Flower People* itself. Solecki acted as the scientific advisor of Constable's book and wrote a brief introduction for it. Unsurprisingly the Shanidar finds, in particular the flower burial, feature prominently in it.[43] *The Neanderthals* was part of "The Emergence of Man Series" and had the internationally operating distribution machinery of the Time Life conglomerate behind it. In the very same year 1973 the book was translated into German, French and Dutch (1973), and some years later into Spanish (1975) and Italian (1979). A considerable numbers of subsequent editions followed until the 1990s.

5. Jean Auel and The Clan of the Cave Bear

Among the most avid readers of the books of both Solecki and Constable was US author Jean Auel (*1936). [44] Since the 1980s she is possibly the best-known writer of palaeofiction. Between 1980 and 2011 Auel published the six volumes

42   Pfeiffer, John E., The Emergence of Man, 3rd ed., New York: Harper & Row, 1978, 155. Pfeiffer, John E., The Creative Explosion: An Inquiry into the Origins of Art and Religion, New York: Harper & Row, 1982, 99. Trinkaus and Shipman, The Neandertals, 340–341. Stringer, Chris and Clive Gamble, In Search of the Neanderthals: Solving the Puzzle of Human Origins, New York: Thames and Hudson, 1993, 98, 158;.Tattersall, Ian, and Jeffrey H. Schwartz, Extinct Humans, New York: Westview, 2000, 215–217. With specific reference to the flower burial: Tattersall, Ian, Becoming Human. Evolution and Human Uniqueness, San Diego/New York/London: Harcourt Brace & Company, 1998, 161, 229. Tattersall, Ian, The Monkey in the Mirror. Essays on Science of What Makes Us Human, San Diego/New York /London: Harcourt, 2002, 124–125. Schrenk and Müller, Die Neandertaler, 80, 98. Shreeve, The Neandertal Enigma, 53.

43   Constable, George, The Neanderthals, The Emergence of Man, New York: Time Life, 1973. I owe this hint to Chris Stringer.

44   Personal information Jean Auel.

of her Earth's Children series. Her novels were translated into over thirty languages and sold worldwide over 45 million copies.

The importance of the Shanidar Neanderthals for Auel's first book, *The Clan of the Cave Bear* is well known and already tangible in the title. The clan's cave resembles the Shanidar cave although Auel put it on the Crimea.[45] The book tells the story of how Ayla, the female hero of the saga, a *Homo sapiens*, grows up within a group of Neanderthals. The character of Creb, the disfigured and handicapped shaman of the group and a father figure for Ayla is modelled on Shanidar I.[46] And indeed, Creb is killed by an earthquake, just like Shanidar I (and not buried).[47] Iza is the medicine woman, knowledgeable about the healing properties of plants, and a mother figure for Ayla. When she dies at the end of *The Clan of the Cave Bear*, Ayla goes out collecting a large number of different plants so Iza would have her "tools" in the next world as well. This way the flower burial appears not as a customary Neanderthal practice but as an exception, conceived of by a *Homo sapiens*.

Auel's Neanderthals are portrayed as human, in the spirit of Solecki, one might say. But they lack mental mobility and do not evolve. Their ability to speak is severely limited, reflecting the scientific debate on Neanderthal speech around 1980.[48] Yet, and here Auel introduces an entirely new dimension, *The Clan of the Cave Bear* can also be read as a work of feminist fiction. In the strictly patriarchic Neanderthal clan females are totally subdued. Ayla is regularly raped by her arch-enemy Broud, the son of the clan leader and gives birth to a boy. (As mentioned above, one recurring topic in paleofiction is the intriguing question of "interbreeding" between the two human species.) She

---

45   There are websites that reconstruct to which Paleolithic sites Auel alludes to in her novel, see e.g. https://www.donsmaps.com/indexauelfans.html.

46   This has been observed already by numerous authors, e.g., Sommer, The Neandertals, 162. Trinkaus and Shipman, The Neandertals, 341. José María Bermúdez de Castro, "Próximo Oriente: un agujero negro de la prehistoria," Quo, 14.5.2015, http://reflexiones-de-un-primate.blogs.quo.es/2015/05/14/proximo-oriente-un-agujero-negro-de-la-prehistoria. Jean Auel said so in interviews, too: e.g., Ken Ringle, "Jean M. Auel: The Smashing Saga of the 'Cave' Woman," The Washington Post, 21 February 1986. "Interview with Jean M. Auel", Goodreads, published electronically April, 2011, https://www.goodreads.com/interviews/show/580.Jean_M_Auel.

47   Auel, Jean M., The Clan of the Cave Bear, New York: Crown, 2011 [1980], 489–490.

48   De Paolo, Charles, Human Prehistory in Fiction, Jefferson, N.C./London: McFarland & Co, 2002, 113–116; Ruddick, The Fire in the Stone, 85.

is also excluded from using weapons and participating in the hunt. Ayla is able to overcome these restrictions but is eventually expelled from the clan.[49]

In the appropriation of Solecki's ideas in paleofiction the comparison of, or shall we rather say the contrast between Neanderthals and "us", represented by Ayla, is explicit. The narrative acts out these differences. In a double movement the Neanderthals appear as human, empathetic and solidary creatures, yet at the same time they are kept at bay. With specific reference to paleofiction and Auel's novels paleoanthropologists Milford Wolpoff and Rachel Caspari write: Neanderthals "became more like the natives whose roles they inherited by becoming 'the other'. ... Neandertals provided the clearest picture of what we were not, and perhaps some insight into how we came to be successful at their expense."[50]

Auel has a reputation of being widely read in prehistory, keeping up with the latest research. Her novels reflect current scientific debates, for example as regards the linguistic abilities of the Neanderthals, or the disk-core technique, i.e. how to knap stone tools.

The collaboration between prehistorians and novelists in producing paleofiction can be traced back until the early twentieth century. Aspiring authors turned to reputed scholars for advise and asked for public endorsement in the form of forewords. In the twentieth century, for example, Henry Fairfield Osborn, the powerful president of the American Museum of Natural History in New York, hoped to push his scientific ideas but also his political agenda by supporting writers of paleofiction.[51] Instances of this kind of collaboration can be found later well into the twenty-first century. On occasion the researchers themselves write paleofiction. The best-known example is the paleontologist Björn Kurtén, but also highly regarded paleoanthropologists such as Juan Luis Arsuaga, François Bordes, Yves Coppens and Donald Johanson dabbled with this genre.[52]

---

49   On Auel's paleofiction see De Paolo, Human Prehistory, 113–119. Ruddick, The Fire in the Stone, 84–89, 179ff.

50   Wolpoff, Milford H., and Rachel Caspari, Race and Human Evolution. A Fatal Attraction, New York: Simon & Schuster, 1996, 277.

51   Sommer, Marianne, History Within. The Science, Culture, and Politics of Bones, Organisms, and Molecules, Chicago: The University of Chicago Press, 2016, 112–120.

52   Hochadel, Oliver, "Ursprung und Überwindung. Heldengeschichten aus Atapuerca." In Erzählung und Geltung. Wissenschaft zwischen Autorschaft und Autorität, edited by S. Azzouni, S. Böschen, and C. Reinhardt, Weilerswist: Velbrück Wissenschaft, 2015,

The relationship between the writer of paleofiction and the professional paleoanthropologist could be described as one of mutual benefit: public visibility in exchange for scientific credibility. In most cases, this relationship is asymmetrical, i.e. the writers depend on the endorsement of the scientists. The case of Jean Auel might be different. Due to her enormous success it seems that the roles are nearly reversed. That is to say that researchers may need her more than the other way round. Prehistorians praise her unflagging curiosity and in-depth knowledge of the subject matter. In their view Auel significantly helped their own cause by creating a sustained interest in prehistory among millions of readers worldwide.[53]

Her economic success with the Earth's Children series allows Auel to support Paleolithic research financially, both conferences and excavations, e.g. in Spain. A photo from a 1986 conference in Santa Fe reunites the crème-de-la-crème of Neanderthal researchers (Lewis Binford, Jean-Philippe Rigaud, Chris Stringer, Erik Trinkaus and Milford Wolpoff, among others) and, as the captions informs, "Jean Auel (a conference sponsor)".[54]

When the final volume of Auel's saga appeared in 2011, first-rate paleoanthropologists such as Ian Tattersall or Chris Stringer presented *The Land of the Painted Caves* to large audiences.[55] French prehistorian Jean-Philippe Rigaud even published a small book on the occasion: *Le Monde des Enfants de la Terre*, an up-to-date guide to every-day-life in the Paleolithic documenting Auel's in-depth knowledge.[56]

In their popular science books on Neanderthals paleoanthropologists comment on what Auel's paleofiction might contribute to the scientific debate, e.g. on gender issues.[57] Archaeologist John Speth points out that paleofiction contains answers to questions relevant to his own discipline: Is it possible to boil liquids, a broth for example, without the use of heated stones? As Jean Auel "explains" in *The Clan of the Cave Bear* one may use perishable materials such as hide to cook. Even if the hide is directly put into the flame

107–132, 127. Hochadel, Oliver, "Spain's Magical Mountain. Narrating Prehistory at Atapuerca," The British Journal for the History of Science 49 (3), 2016: 453–472, 466–467.

53   Ruddick, The Fire in the Stone, 87–88.
54   Stringer and Gamble, In Search of the Neanderthals, 191.
55   Diario de Atapuerca 39 (2011), 10 (Tattersall on Auel).
56   Rigaud, Jean-Philippe, Le monde des enfants de la terre. Comment vivaient les héros de la saga de Jean M. Auel, Paris : OMNIBUS, 2011.
57   Stringer and Gamble, In Search of the Neanderthals, 33.

it will not burn as long as there is water in it.[58] US archaeologist Lawrence G. Straus collaborated with Jean Auel for many years, for example arranging visits to prehistoric sites for her. Straus is convinced that prehistorians may learn a lot from writers of paleofiction such as Auel or science writers such as John Pfeiffer providing illuminating syntheses of research.[59]

6.   Imagining the Flower Burial

Hollywood also took note of the enormous success of Auel's paleofiction. In 1986 the movie *The Clan of the Cave Bear* directed by Michael Chapman and starring Daryl Hannah as Ayla was shown on the big screen. The characters communicate in some rudimentary language made comprehensible through subtitles. Two key features from the original Solecki narrative in Auel's adaptation reappear: the severely handicapped shaman Crab and the flower burial of the medicine woman Iza. The flower burial is only a brief scene and unlike in the novel Ayla is not shown as the one collecting the flowers.[60]

The movie is considered a failure, doing well neither at the box-office nor with film critics who thought it predictable and unoriginal. As Roger Ebert put it: "the actors are asked to play characters who are modern in everything but dress and language. ... There is no sense of the alien and the unknown, no sense that these people have ideas and feelings that would be strange to us."[61] One film critic (in a much later review from 2012) even perceived a racist or eugenic undertone. Blond and blue-eyed Ayla simply excels in everything she is doing (Ebert had dubbed her "the first Rhodes Scholar") while the Neanderthals are portrayed as unable to evolve and thus, we may infer, were doomed to go extinct.[62]

---

58   Speth, John D., "When Did Humans Learn to Boil?," PaleoAnthropology, 2015: 54–67, 57. This method is common knowledge, as Speth points out, yet only Auel, The Clan, 81, makes the reference to prehistory.

59   Straus helped Pfeiffer gathering material for his book The Creative Explosion.

60   Michael Chapman, "The Clan of the Cave Bear," (Warner Bros., 1986), 98 minutes. On the movie see Ruddick, The Fire in the Stone, 91–92.

61   Ebert, Roger, "The Clan of the Cave Bear," Chicago Sun-Times, 21 February 1986; also see Maslin, Janet, "Screen: 'Clan of the Cave Bear'," The New York Times, 17 January 1986.

62   Brayton, Tim, "Blockbuster History: Life in the Pleistocene", published electronically 16 July 2012, https://www.alternateending.com/2012/07/blockbuster-history-life-in-the-pleistocene.html.

Jean Auel herself hated the adaptation because she felt that the cinematic adaptation of her first novel was highly inaccurate and too violent. She sued the producers for 40 million dollars.[63] Although initially planned, no second part was made. Just a few years ago there was a new attempt to turn Auel's saga of *The Clan of the Cave Bear* into a TV series. A pilot was shot but never screened and the project ended stillborn.[64]

The crucial importance of the visualization of early humans for both, the general public but also for the researchers themselves, has been stressed repeatedly. Drawings, dioramas and full-body-reconstructions are "influential documents which play a part in the shaping of archaeological debates".[65] The strong impact of the flower burial on popular culture is also owed to the numerous visual representations of it. The moving story required a moving image.

Solecki's book is well illustrated with photos and drawings but none of these show the flower burial. Constable's book from 1973 contains the first visual depiction of the flower burial in the form of a large sketch across two pages by Herb Steinberg. *Origins* (1977), the bestseller by Richard Leakey already mentioned, features another colorful rendering of the flower burial by US artist Ronald Bowen, also spread over two pages. For Leakey the flower burial is a fact, supporting his overall argument of the social and cooperative nature of human beings (including their ancestral species).[66]

Many of the representations in 2D (books, magazines, video animations) and 3D (dioramas in museums) accentuate the variety of colors of the flowers and thus the inherent beauty of the act. What all the images have in common is the solemnity of the burial, the seriousness of the Neanderthals bidding farewell to the deceased member of their group. On some of the images the mourning clan members embrace as each other in search for comfort. The message is the same: these prehistoric beings are indeed fully human.

63   Ringle, "Jean M. Auel".
64   https://en.wikipedia.org/wiki/The_Clan_of_the_Cave_Bear#Film_and_television_ada ptations (last accessed 5 February 2019) and personal communication by Lawrence G. Straus and Jean Auel.
65   Moser, Stephanie, "The Visual Language of Archaeology: A Case Study of the Neanderthals," Antiquity 66, 1992: 831–844, 831. Similarly Moser, Stephanie, "The Dilemma of Didactic Displays. Habitats Dioramas, Life-Groups and Reconstructions of the Past." In Making Early Histories in Museums, edited by N. Merriman, London: Cassell, 1999, 95–116, 105.
66   Leakey and Lewin, Origins, 126–127.

7. Serious doubts about the Flower Burial

"Solecki's view of Neandertals as human, humane, compassionate, and caring was accepted widely and with remarkably little demur ... Neandertals were simply flower children under the skin. The skepticism came later", wrote Erik Trinkaus in 1993.[67] In the meantime, the theory of the flower burial by Solecki and Leroi-Gourhan is considered erroneous by most researchers. There has been a number of alternative explanations for the large number of pollen in the soil samples.

The first substantial criticism was published in 1989 in an article that questioned intentional Neanderthal burial *tout court*. With respect to the Shanidar flower burial its author Robert Gargett alleged that wind or animals were the far more likely agents to have introduced the pollen into the sediments of the cave, accusing Solecki and Leroi-Gourhan more or less explicitly of wishful thinking.[68]

In 1999, Jeffrey Sommer suggested that rodents actually were responsible for the high concentration of pollen found by Solecki.[69] Sommer referred to the work of Richard Redding who had excavated a number of *Meriones crassus* and *Meriones persicus* (two species of Persian jirds, related to gerbils) from the Zagros mountains, the region where also the Shanidar cave is located. These rodents pack their burrows with flower heads and seeds. This would explain the high number of pollen found by Leroi-Gourhan. *Meriones persicus* is known to burrow down several meters and occupy the areas in front of caves as they love the softer soil.[70]

The lack of access to the site until a few years ago made it difficult to resolve the issue. Yet in recent years, excavations resumed and it was shown that the sediments of Shanidar cave were indeed full of burrows. In 2015, a "transect" of surface samples from inside and outside the cave was collected and screened for pollen. The researchers suggested that bee-carried pollen and wind-blown vegetation that had been covered up over the last 60,000 years were responsible for Leroi-Gourhan's findings.[71]

---

67   Trinkaus and Shipman, The Neandertals, 341.
68   Gargett, "Grave Shortcomings", 176.
69   Sommer, Jeffrey D., "The Shanidar IV 'Flower Burial': A Re-Evaluation of Neandertal Burial Ritual," Cambridge Archaeological Journal 9 (1), 1999: 127–129.
70   I thank Richard Redding for this information.
71   Fiacconi, Marta, and Chris O. Hunt, "Pollen Taphonomy at Shanidar Cave (Kurdish Iraq): An Initial Evaluation," Review of Palaeobotany and Palynology 223, 2015: 87–93.

When confronted with the evidence questioning the flower burial in 2004, Auel said: "That may be the case. But it's nevertheless a good story."[72] It nearly seems that the "good story" of the flower burial needs to be refuted over and over again, both in the academic and in the public sphere. The Neandertal Museum in Mettmann (Germany) for example does not display the flower burial, but tongue-in-cheek shows a photo of *Meriones persicus*. The caption of the photo explains that the rodent presumably introduced the pollen into the grave.[73]

Nevertheless the flower burial is still on display in a number of museums such as the Gunma Museum of Natural History in Japan, that opened in 1996.[74] Ian Tattersall acted as scientific advisor for the artist Shuichiro Narasaki. Nowadays the curators of the Gunma Museum are of course aware of the fundamental critique of the hypothesis of the flower burial which they share. In their tours of the exhibit they explain to the visitors that the diorama does not represent any more the scientific consensus but for the moment they will maintain the diorama.[75]

The persistence of the story of the flower burial and its visual representation might be attributed to the powerful image it conveys (which makes it attractive to museum curators.) As far as museum exhibits are concerned, it also has to do with the inherent conservatism and thus long lives of these costly exhibits.[76]

As recently as 2010 a representation of the flower burial was put on display in the David-H.-Koch-Hall-of-Human-Origins of the Smithsonian National Museum of Natural History in Washington.[77] The diorama was produced by artist Karen Carr under the scientific supervision of Richard Potts and Briana Pobiner. The two paleoanthropologists from the Smithsonian maintain that the hypothesis of Leroi-Gourhan can be defended against the criticism. The Smithsonian was one of the main sponsors of the excavations in the 1950s, so in a sense this exhibit on Shanidar documents its success. By the way:

---

72   Husemann, Dirk, Die Neandertaler. Genies der Eiszeit, Frankfurt a.M.: Campus, 2005, 20.

73   Personal communication, Bärbel Auffermann, Neandertal Museum Mettmann.

74   The diorama is reprinted in Tattersall and Schwartz, Extinct Humans, 216.

75   Personal communication.

76   Moser, "The Dilemma," 111.

77   Edwards, Owen, "The Skeletons of Shanidar Cave," Smithsonian Magazine, March 2010. http://www.smithsonianmag.com/arts-culture/the-skeletons-of-shanidar-cave-7028477/?no-ist=

The Smithsonian is the only institution outside Iraq that holds some of the Neanderthal fossils found by Solecki and his team: Shanidar III. This partial skeleton has a different story to tell: a rib injury caused by a thrown spear. Thus Shanidar III is on occasion labeled the "first murder".[78] This constitutes another potential story-line with popular appeal, addressing the issue of interpersonal violence and its long prehistory.[79]

While the flower burial of Shanidar IV has been by and large discarded the story of Shanidar I has rather gained momentum in recent years. In a recent study Trinkaus and Villotte showed that Shanidar I also must have had severe hearing problems making him even more vulnerable (and thus in need of even more protection from the members of his group).[80] The interpretation that this individual was severely handicapped and could only reach an advanced age because of the continuous help from member of his group as such has never been questioned. Rather Shanidar I has become a crucial reference in showing that early humans already felt and practiced compassion. Compassion and related concepts such as empathy, altruism and social cohesion have become a "trending topic" in the field of prehistory in recent years.[81]

In the meantime, other "famous invalids", individuals with severe bodily (and also mental) disabilities have joined Shanidar I. Some of these were far older than the Neanderthal from Shanidar cave. To give but two examples: In 2005, David Lordkipanizde and his team published a paper on a skull found at the site of Dmanisi in Georgia, dated at around 1.8 million years. The individual had only one tooth left and had lived for many years after the loss of his teeth. The researchers thus proposed that his group must have "prechewed" his food for him.[82]

---

78   Churchill, Steven E. et al., "Shanidar 3 Neandertal Rib Puncture Wound and Paleolithic Weaponry," Journal of Human Evolution 57 (2), 2009: 163–178.

79   For an earlier period see Sala, Nohemi et al., "Lethal Interpersonal Violence in the Middle Pleistocene," PLoS ONE 10 (5), 2015: e0126589.

80   Trinkaus, Erik, and Sébastien Villotte, "External Auditory Exostoses and Hearing Loss in the Shanidar 1 Neandertal," PLoS ONE 12 (10), 2017: e0186684.

81   For a comprehensive synthesis going well beyond HOR see Spikins, Penny A., How Compassion Made Us Human. The Evolutionary Origins of Tenderness, Trust and Morality, Havertown: Pen & Sword Books Ltd, 2015. Critical discussion in Spikins, Penny et al., "Calculated or Caring? Neanderthal Healthcare in Social Context," World Archaeology 50 (3), 2018: 384-403.

82   Lordkipanidze, David et al., "The Earliest Toothless Hominin Skull," Nature 434, 2005: 717–718.

In 2009, the Atapuerca research team published the skull of a young individual from the Sima de los Huesos in Northern Spain, approximately 400,000 years old.[83] It suffered from a rare disorder called craniosynostosis that causes a premature fusion of one of the sutures of the skull and hence deformed it in the growth process. Not only was her face disfigured, it is also most likely that she was developmentally disabled. The researchers argued that group members must have cared deeply for the little creature—otherwise it would have never reached the age of approximately ten years. To make their interpretation palpable they baptized the individual Benjamina, the "loved child". Lead researcher Ana Gracia admitted that there is no way of proving that the group members actually loved her, or even that the individual was actually a female.[84] As part of the press release of their scientific article, the Atapuerca research team provided a reconstruction of Benjamina's presumed appearance. The intention to tell a moving story of a handicapped but beloved girl proved highly successful judging from the enormous echo of the discovery in the Spanish press. The central message of the coverage was that already back then human beings cared deeply for the members of their social group.[85] In a sense, they were already fully human—an argument very much in line with Solecki's *Shanidar. The First Fower People*.

8. Conclusion

That the field of HOR is characterized by specific narrative structures was first forcefully suggested by Misia Landau. In her book *Narratives of Human Evolution*, she argues that in these narratives human evolution (within a given variation of possible sequences) is told as the long and arduous but ultimately successful and heroic journey of "man" becoming "himself".[86] At the center of Landau's study are two authors from the first third of the twentieth century, Arthur Keith and Grafton Elliot Smith. Their interest was primarily of phylo-

83   Gracia, Ana et al., "Craniosynostosis in the Middle Pleistocene Human Cranium 14 from the Sima de los Huesos, Atapuerca, Spain," Proceedings of the National Academy of Sciences of the United States of America 106, 2009: 6573–6578.
84   Corbella, Josep, "Ley de dependencia en Atapuerca," La Vanguardia, 31 March 2009, 25.
85   Hochadel, "Spain's Magical Mountain," 469–470.
86   Landau, Misia, Narratives of Human Evolution, New Haven/London: Yale University Press, 1991.

genetic nature. In their (implicitly teleological) schemes of human evolution "man" is rather a general, an abstract entity.

The narratives of Ralph Solecki (and later on of Donald Johanson and many others) in their popular science books are of a different nature. Their protagonists are concrete individuals: Shanidar I, Shanidar IV or Lucy. The stories of these creatures—as they were deciphered by the paleoanthropologists, archaeologists and palynologists—were written onto their fossilized remains or found in the pollen around them.

The focus on individuals fundamentally changes the story-telling in HOR. Taking Solecki's *Shanidar. The First Flower People* as point of departure this article has traced the productive interaction between archaeological research and popular culture. From the flower burial (Shanidar IV) and the severely handicapped old man (Shanidar I) a rich cosmos of intriguing narratives and powerful images emerged. These were adapted and transformed to be used in different media and for different audiences. The paleofiction of Jean Auel, in particular her first novel *The Clan of the Cave Bear*, was key in spreading these stories. New elements or story-lines were added, for example the gender issue through the character of the female hero Ayla. More recently the topics compassion and empathy have gained prominence both in the academic and the popular sphere showing once more how closely these two spheres are interconnected. Other narratives got lost. In George Constable's book of 1973, the Iraqi Kurds are still present through Solecki's photos that were reproduced. But in the subsequent adaptations and streamlining of the narrative the substantial anthropological part of *Shanidar. The First Flower People* disappeared completely.

Narratives do not appear out of nothing but are often antitheses to already existing and thus competing story-lines. In HOR they address fundamental questions such as human nature. As we have seen, Solecki's proposal of deeply humane Neanderthals was meant to reverse the image of a brute and primitive creature that prevailed at the time. In a much more implicit way Solecki (and later on other researchers such as Richard Leakey in *Origins*[87]) also argued against the idea of man as inherently aggressive and prone to kill. This notion, very much a child of the immediate post-WWII-era, was forcefully

---

87   Ruddick, The Fire in the Stone, 85.

proposed by Raymond Dart in the 1950s, and also strongly reverberated in the popular sphere, dubbed the killer-ape theory.[88]

Although phylogenetic questions remain important the research focus in HOR has widened considerably since the mid-twentieth century. Reconstructing the behavior, the social world including the natural environment of our ancestors have become the main focus point of prehistoric research. In order to answer these new questions HOR became a highly interdisciplinary field, starting in the 1950s and 1960s.

It was not least these changes in the disciplinary set-up of HOR that allowed for the construction of new stories. The crucial role of palynology in the genesis of the flower burial story illustrates this very well (but see below). As mentioned earlier in this article: More recently the analysis of ancient DNA provides new narratives: the Neanderthal lives now in us.[89] These stories are again quite different to the ones emerging from fossils and pollen.[90] In narrative terms this implies a change of the protagonist. These stories are again more "universal" reconstructing the fates of entire populations or species. But once more the driving question is to find the differences between Neanderthals and ourselves, i.e. what makes us distinct. The latest attempt in this respect is the growing of "miniature brains" from Neanderthal DNA in order to detect significant differences in the brain structure between us and them.[91] The enormous narrative potential of these most recent advances in paleogenetics for paleofiction is obvious.[92]

---

88   See e.g. Weidman, Nadine, "Popularizing the Ancestry of Man: Robert Ardrey and the Killer Instinct," Isis 102 (2), 2011: 269–299.

89   Bösl, Elsbeth, Doing Ancient DNA: Zur Wissenschaftsgeschichte der aDNA-Forschung, Bielefeld: transcript, 2017. Dobson Jones, Elizabeth, The search for ancient DNA. In the media limelight: A case study of celebrity science, PhD, University College London, 2017.

90   For the narrative potential of these "genetic stories" more broadly, in particular with respect to the field of population genetics see the work of Sommer, Marianne, "History in the Gene: Negotiations between Molecular and Organismal Anthropology," Journal for the History of Biology 41 (3), 2008: 473–528, 476, and Sommer, History Within, chapter 14: "The Genographic Network: Science, Markets, and Genetic Narratives."

91   Cohen, Jon, "Exclusive: Neanderthal 'Minibrains' Grown in Dish," Science News, 20 June 2018; http://www.sciencemag.org/news/2018/06/exclusive-neanderthal-mini-brains-grown-dish.

92   See for example the novel by Cameron, Claire, The Last Neanderthal, New York: Little, Brown and Company, 2017.

In all these story lines the comparison of Neanderthals and *Homo sapiens* is key. The quest of Solecki and other researchers to humanize Neanderthals renders the question in how far we differ from them even more pressing. The two species can only be understood as a *histoire croisée*. As paleoanthropologists such as Milford Wolpoff and Rachel Caspari succinctly put it: Neanderthals "are at the root of how Western science defines humanity; they are the 'other' to which humanity is compared: they are what a 'modern human' is not."[93] Paleofiction as a genre may be particularly apt in this respect, because it is at complete liberty to "analyze" the differences between both human species.

Clearly, there are more stories out there about the Shanidar Neanderthals, on which this article barely touched upon. One of these stories is an "internal" one and concerns the status of palynology within the field of HOR. The discipline of palynology originated in the early twentieth century in order to analyse the deposition of pollen in specific environments. Yet until the work of Leroi-Gourhan it had been applied only to "natural environments", e.g. dried-out lakes. Her seminal paper on the pollen found in the sediments of Shanidar IV in 1968 dealt with a "human environment", the Shanidar cave. The impact of the paper that led to Solecki's hypothesis of the flower burial was considerable, also because it raised the status of the palynology as a discipline by showing its potential in reconstructing the prehistoric world. Arlette Leroi-Gourhan created her own school at her laboratory at the Musée de l'Homme in Paris. Yet her approach to analyse the sediments of caves (in Shanidar and elsewhere) has been severely criticized mostly by Anglo-American scholars.[94] Because human agency severely disturbs the sediments containing the pollen the method may only applied, as before, to natural, i.e. untouched environments. Yet the jury of what palynology might be able to achieve and what not is still out there. Among Leroi-Gourhan's mostly French students the flower

---

93   Wolpoff and Caspari, Race and Human Evolution, 270; already quoted by De Paolo, Human Prehistory, 144. Many historians of science have made the same point: see Sommer, The Neandertals, 139–140. Schweighöfer, Ellinor, Vom Neandertal nach Afrika. Der Streit um den Ursprung der Menschheit im 19. und 20. Jahrhundert, Göttingen: Wallstein, 2018, 119, 130, 140.

94   E.g. Turner, C. and G.E. Hannon, "Vegetational Evidence for Late Quaternary Climatic Changes in Southwest Europe in Relation to the Influence of the North Atlantic Ocean," Philosophical Transactions of the Royal Society of London B 318, 1988: 451–485.

burial is still cited as a towering achievement in articles that were published years after the fundamental criticisms cited above.[95]

Another issue this article did not pursue is the appropriation of the finds among the Kurdish population of the region. Graeme Barker, an archaeologist of the University of Cambridge, excavating in Shanidar since 2014, said in an interview: "... in Kurdish society there is general awareness that they have this incredibly important, very early site with Neanderthals. They're very proud of Shanidar. A local society a few years ago put money into making the site accessible: it has got a Peshmerga guard, car parking, information panels, steps up to the site, and picnic places. When the weather is good in the spring, on Fridays and Saturdays there are usually large numbers of Kurdish visitors."[96] A bust of Ralph Solecki outside the cave commemorates the discoverer of Shanidar.

In January 2019, at the University of Cambridge, British and Iraqi researchers met at a symposium entitled "Neanderthal Notions of Death and its Aftermath: The Contribution of New Data from Shanidar Cave". Surely, there are more stories about the Shanidar Neanderthals in store.

Acknowledgements: I would like to thank Jean Auel, Bärbel Auffermann, Juanjo Ibáñez, Dishad A. Mutalb, Montserrat Obiols, Briana Pobiner, Emma Pomeroy, Richard Redding, Penny Spikins, Lawrence G. Straus, Chris Stringer, Ian Tattersall and Seigo Yamashina, for providing me with valuable information. Research for this article was supported by the Grup de recerca consolidate I finançat (2017 SGR 1138, AGAUR-Generalitat de Catalunya).

95   Bui, Thi Mai, and Michel Girard, "Pollens, ultimes indices de pratiques funéraires évanouies," Revue archéologique de Picardie NS 21, 2003 : 127–137, 128.

96   Callaway, Ewen, "Archaeologists Ousted by Isis Return to Ancient Iraqi Cave," Nature online, 2 October 2015, https://www.nature.com/news/archaeologists-ousted-by-isis-return-to-ancient-iraqi-cave-1.18487

Narrating and Comparing in the Organization of Research Projects

Rebecca Mertens

Comparisons are ubiquitous in the history of political thought, in the arts and in the sciences. They seem crucial to our understanding of the cultural and natural world as well as to the establishment of social norms.[1] Doing comparison is here understood as an activity which is always situated within cultural practices, enforcing certain ways of comparing and suppressing others. At the same time, these practices lead to the dominance of particular comparisons over others and may thereby strongly shape social discourses, as the recent example of public communication in the case of the coronavirus pandemic shows.[2] Studying how doing comparisons influences social life thus means to gain more information about the processes which are initiated, altered or stabilized by the use of comparison in particular contexts, e.g. in political conflicts, in the development of economic markets, or in the making and professionalization of art and science. The study of the practices of comparison should also allow us to gain further knowledge about the ways in which

---

1   Epple, Angelika, and Walter Erhart, "Die Welt beobachten. Praktiken des Vergleichens." In *Die Welt beobachten. Praktiken des Vergleichens*, edited by A. Epple and W. Erhardt, Frankfurt a.M.: Campus, 2015, 7f.

2   Apart from daily comparisons between the dimension of the pandemic in different countries, including infection and mortality rates, the coronavirus pandemic has recently been compared to the 1918 flu pandemic in the German public press. Similar to former public displays of pandemic prevention in Germany and in the US, this historical comparison has been used to communicate future risks and the necessity of prevention policies. See i.e. Der Spiegel "Wenn die zweite Welle kommt," 24.04.2020; Tagesspiegel, "Die Mutter der modernen Pandemien. Coronavirus und Spanische Grippe im Vergleich," 20.03.2020. For a detailed analysis of the role of historical comparisons and narratives in pandemic prevention discourses, see Rengeling, David, *Vom geduldigen Ausharren zur allumfassenden Prävention. Grippe-Pandemien im Spiegel von Wissenschaft, Politik und Öffentlichkeit*, Baden-Baden: Nomos, 2017.

comparing is related to other social, epistemic or cultural activities like social networking, the introduction of new cultural and scientific techniques,
as well as categorization and standardization, to name just a few. This paper contributes to the explication and historical reconstruction of practices
of comparison in the mid-20th century by analyzing how doing comparison
has affected the social structure and development of research projects in the
context of the biochemical sciences in the United States. So far, the role of
comparison in scientific project organization is more or less a blind spot in
the history and philosophy of science.

The role of comparison in project organization and development will
be examined by means of a historical case study on collaborative research
projects that were conducted in the 1940s and 50s between the departments
of biology and chemistry at the California Institute of Technology (Caltech).
Throughout the study, it will become clear that the comparisons made in
the realm of project organization were placed within a particular semantic
and narrative context, created in the process of proposal and report writing,
professional correspondence between scientists and research managers and
science popularizing activities. In fact, I will argue that comparisons drawn
between different research areas and their objects of study at Caltech gained
their validation and their persuasive power from *project organization narratives*.
This case, then, gives us the possibility to locate practices of comparing and
narrating in scientific and science policy discourses in the US during the
Second World War and throughout the Cold War period. What is more,
the case shows that scientific popularization, public relations work and
grant management are in fact closely related practices which provide and
instantiate instruments and narratives for the development of research in
the 20th century.[3]

---

3   Recent studies in the philosophy and sociology of scientific practice support the idea
that the success and relevance of cross-disciplinary research projects depend in many
ways on the ability of scientists to perform a variety of activities across different contexts and to connect them. These studies focus on the connections of different practices in the development of scientific projects, programs, disciplines and infrastruc-
tures, studying the dynamics these interrelations of practices invoke on scientific and
social change. For the cognitive and social affordances on interdisciplinary science organization and change, see e.g., Andersen, Hanne, "Collaboration, interdisciplinarity,
and the epistemology of contemporary science," *Studies in History and Philosophy of Science* 56, 2016: 1-10, as well as Nersessian, Nancy J., and McLeod, Miles, "Interdisciplinary
problem-solving: emerging modes in integrative system biology," *European Journal for*

1. Funding strategies and research cooperation in the early 20th century

Philanthropic foundations played a central role in the growth of large-scale research programs in the US in the first half of the 20th century. Especially the Rockefeller Foundation (RF) had its share in the development of the experimental sciences and in the rising popularity of molecular approaches in biology and biomedicine.[4] During the 1920s and early 1930s, the RF re-organized its "system of patronage,"[5] focusing on individual project grants in selected scientific areas rather than on entire institutions. Projects which qualified for support were supposed to contribute "in a basic and important way to mankind," they further had to be "sufficiently developed to merit support, but so imperfectly developed as to need it."[6] By applying these selection criteria, the RF hoped to "play a critical role in producing and stimulating development that otherwise would not occur within reasonable time."[7] Another aspect of the new system was the authority and responsibility of the division officers over the general program, the choice of research topics, and individual grantees. This required experience and training in the respective fields

---

Philosophy of Science 6(3), 2016: 401-418. The role of knowledge implementation in the expansion and "re-situation" of research projects has recently been elucidated by Meunier (See Meunier, Robert, "Project Knowledge and its Re-situation in the Design of Research Projects," *Studies in History and Philosophy of Science Part A*, 2018, ISSN: 0039-3681, accessed May 2020, https://www.sciencedirect.com/journal/studies-in-history-and-philosophy-of-science-part-a/articles-in-press). Ankeny and Leonelli introduced the term of "research repertoires" to grasp the social and epistemic activities involved in successful research development and organization in the late 20th century. Such "repertoires" can be characterized as ways of making intended research performances work, including a very diverse set of material as well as conceptual components and activities, and most importantly the ability of scientists and science managers to successfully combine them. See Ankeny, Rachel, and Leonelli, Sabrina, "Repertoires: A post-Kuhnian perspective on scientific change and collaborative research," *History and Philosophy of Science* 60, 2016: 18-28.

4   Kohler, Robert E., *Partners in Science. Foundations and Natural Scientists, 1900-1945*, Chicago: The University of Chicago Press, 1992, 233-303.

5   Kohler, *Partners in Science*, 231.

6   Rockefeller Foundation. *The Natural and Medical Science Cooperative Program*, December 13, 1933, 4, 100 Years: The Rockefeller Foundation, accessed May 4, 2020, https://rockfound.rockarch.org/digital-library-listing/-/asset_publisher/yYxpQfeI4W8N/content/the-natural-and-medical-sciences-cooperative-program.

7   Rockefeller Foundation. *The Natural and Medical Science*, 4f.

of research as much as the ability to identify potential scientific trends and communities.[8] Warren Weaver, a former mathematician who became head of the RF's natural science division in 1932, used these new possibilities to promote physical and chemical approaches in application to biology.[9] Shortly after Weaver joined the RF as their natural science officer, the division settled a new program, focusing on the relations between biology and the medical sciences as well as on behavioral biology and psychology. The "Science of Man," as Weaver called the new field that ought to be developed by the help of the Foundation, aimed to unlock the basics of human behavior; "man's conceiving, child-bearing, thinking, behaving, growing and finally dying."[10] Key to this approach was the increasing support of research along the lines of neurophysiology as well as research in chemistry and physics in relation to biological problems, including "nutrition, radiation physics, chemical embryology, genetics, physiology, and biochemistry."[11] In the early and mid-20th century, Caltech became one of the most attractive scientific institutes for the RF as part of their funding program in the field of "experimental" and "molecular biology." During this time the RF significantly supported research in biology and chemistry at Caltech, beginning with Thomas Hunt Morgan's work in genetics and Linus Pauling's research on the physico-chemical structure of proteins.[12] Especially in the latter case, the Foundation's agenda in the advancement of interdisciplinary research shaped the general scope of research projects carried out under the supervision of Pauling and his collaborators. As for the implications of these interrelations for the development of molecular biology in the United States, there are many informative and profound historical studies on this episode, including Kay's canonical study on Caltech's role in

---

8   Kohler, *Partners in Science*, 234f.

9   Kohler, *Partners in Science*, 265

10   Weaver, Warren. "The Science of Man," November 29, 1, 100 Years: The Rockefeller Foundation, accessed May 4, 2020, https://rockfound.rockarch.org/digital-library-listing/-/asset_publisher/yYxpQfeI4W8N/content/the-science-of-man. Kay identifies the "genetic control of human behavior" as a dominant theme in the American eugenics discourse of the 1930s. Kay, Lily E., *The Molecular Vision of Life: Caltech, The Rockefeller Foundation, and the Rise of the New Biology*, New York/Oxford: Oxford University Press, 1993, 39.

11   Rockefeller Foundation. *The Natural and Medical Science*, 4.

12   Abir-Am, Pnina, "The Rockefeller Foundation and the Rise of Molecular Biology," *Nature Reviews—Molecular Cell Biology* 3, 2002: 65-70.

the foundational period of molecular biology,[13] as well as Abir-Am's studies on the power structure of the RF,[14] and Robert Kohler's work on the development of philanthropic organizations in the United States.[15] The aim of the present study thus cannot be to re-write these histories of the early beginnings of molecular biology. Rather, I will take a closer look at the ways in which the biochemical research program around Linus Pauling at Caltech was organized by means of comparison, particularly within the writing of proposals and in related practices, such as public-relations work and social networking. It will be shown that the continuing and growing influence of the RF supported a specific way of research collaboration at Caltech that was driven by the use of comparison between the domains of protein chemistry, immunology, embryology, and genetics. I will further argue that the successful use of comparison (between research domains and phenomena) in the context of grant management depended in many ways on the development of what can be called project organization narratives and their oscillation between the context of grant management, public relations and science popularization.[16]

1.1   Research on the structure of biologically significant macromolecules

Linus Pauling's chemistry- and physics-driven research on the structure of biologically significant substances was continuously supported with RF grants

---

13   Kay, Lily E., "Molecular Biology and Pauling's Immunochemistry: A Neglected Dimension," *History and Philosophy of the Life Sciences* 11 (2), 1989: 211-219. See as well Kay, Molecular Vision.

14   Abir-Am, Pnina, "The Biotheoretical Gathering, transdisciplinary authority and the incipient legitimation of molecular biology in the 1930s. New perspectives on the historical sociology of science," *Journal of the History of Science* 25, 03/1987: 1-70. Abir-Am, Pnina, "Molecular Biology in the context of British, French, and American cultures," *International Social Science Journal* 168, 2001: 187-199, and Abir-Am, Pnina, "The Rockefeller Foundation and the Rise."

15   Kohler, *Partners in Science*.

16   Approaching the sociology of science popularization from the perspective of discourse theory, Greg Myers defined popularization as a "continuum" of "different ways of speaking for different rhetorical purposes." In this understanding, popularizing statements and narratives are not bound to specific genres. They actually appear in almost any scientific genre, be it grant proposals, scientific journal articles, research reports, or autobiographies. See Myers, Greg, "Discourse studies of scientific popularization: questioning the boundaries," *Discourse Studies* 5 (2), 2003: 265-279, on p. 270.

during the following decades.[17] At first, Pauling directed his attention to the chemical structure of proteins, such as insulin and hemoglobin, in relation to medical problems.[18] In the late 1930s, he began to work on antibodies and on the mechanism of their formation; this was part of the larger project on proteins of biological and medical relevance. Pauling contributed to the already nourishing field of immunochemistry with his "template theory" of antibody formation. It explained the process by which normal globulins were turned into specific antibodies, that is, into substances that could selectively react with certain antigens and prevent infection.[19] The "template-theory" focused on the complementary relationship between two immunologically effective chemical agents, the antigen and the antibody, and aimed to provide an answer as to how the antibody could react in a complementary manner with the respective antigen in terms of chemical and physical forces. The concept of complementarity hitherto played an important role in late 19th century stereochemistry and immunology, i.e., in Paul Ehrlich's side-chain theory of immunity.[20] In the context of the side-chain theory, however, ideas on complementarity on the molecular level were embedded in a broader framework of normal and pathological cellular processes and linked to assumptions about the origins of antibodies within the living organism. According to this theory, antibodies were formed by receptors formerly involved in normal, that is, non-pathological, cellular processes and would then link themselves to the respective antigen and neutralize its harming effects due to their matching chemical configuration.[21] In the 1920s and 30s, Ehrlich's model of the antibody-antigen reaction was modified and interpreted in the light of new quantitative chemical methods for the study of macromolecules. Amongst others, Michael Heidelberger, Stuart Mudd, Friedrich Breinl, and Felix Haurowitz developed so-called "instructive" theories of antibody-formation, proposing that antibodies were products of a process of chemical transformation, induced

---

17   See Kay, *Molecular Vision*, 143f.
18   See Strasser, Bruno, "Sickle Cell Anemia, a Molecular Disease," *Science* 286 (5444), 1999: 1488-1490, on p. 1488.
19   Pauling, Linus, "A theory of the structure and process of antibody formation," *Journal of the American Chemical Society* 62(10), 1940: 2643-2657.
20   See Cambrosio, Alberto, et al., "Arguing with images: Pauling's Theory of Antibody formation," *Representations* 89(1), 2005: 94-130, on p. 108.
21   Ehrlich, Paul, "On Immunity with Special Reference to Cell Life," *Proceedings of the Royal Society of London* 66, 1900: 424-448.

by the presence of an antigen.[22] Pauling developed this idea further and proposed a detailed stereochemical mechanism by which a chain of normal globulin folded into a structure that was complementary to the antigen; a process
which was initiated by the close physical approximation of the globulin and
the antigen.[23]

Pauling's immunochemical research was especially appealing for the RF
and for Caltech's public image in the 1940s for two reasons: First, it provided
a role model for one of the major goals of the RF with respect to the development of the natural sciences in the US, namely, the successful application of physical methods and theories to biological problems.[24] Secondly, it
was closely related to the possibility and the promise of synthetic antibody
production, which was not only a general concern of human wellbeing and
health but also ultimately connected to war efforts. It is thus not surprising
that the governmental Office of Scientific Research and Development (OSRD),
as well as the RF and other funding organizations with an interest in the advancement of biomedical research, such as the *Foundation for Infantile Paralysis*,
strongly supported the expansion of antibody research throughout the early
and mid-1940s.[25] Yet, as Kay notes, the focus on antibodies and immunological problems was in some way a deviation from the RF's early molecular
biology program in the natural sciences which focused on the application of
physical methods to biology rather than on research with strong medical implications.[26]

On a more theoretical level, Pauling's work on antibodies contributed to
the generalization of claims concerning the physico-chemical interpretation
of specific macromolecular interactions active in living organisms. In several
articles and talks, Pauling argued for the general biological importance of
complementarity and complementary template construction.[27] For instance,

---

22   Silverstein, Arthur M., *A History of Immunology*, Cambridge MA: Academic Press, 1989,
      51f.

23   Linus Pauling et al., "The Nature of the Forces between Antigen and Antibody and of
      the Precipitation Reaction," *Physiological Reviews* 23(3), 1943: 203-219. For an overview
      on Pauling's work in the context of the debate on immunological specificity, see
      Mazumdar, Pauline M.H., *Species and Specificity. An Interpretation of The History of Im-
      munology*, Cambridge MA: Cambridge University Press, 1995, 330ff.

24   Abir-Am, "Molecular Biology in the context," 193.

25   Kay, *Molecular Vision*, 172.

26   Kay, *Molecular Vision*, 164.

27   See, for instance, Pauling, Linus, and Max Delbrück, "The Nature of Intermolecular
      Forces operative in Biological Processes," *Science* 92(2378), 1940: 77-79. Pauling, Linus,

in his 1946 paper on "Molecular architecture and biological processes," Pauling pointed to the "very strong evidence that the specificity of the physiological activity of substances is determined by the size and shape of molecules."[28] The responsible mechanism underlying specific physiological processes was supposedly the same found in the case of antibody formation. Hence, "the size and shape find expression by determining the extent to which certain surface regions of two molecules (at least one of which is usually a protein) can be brought into juxtaposition—that is, the extent to which these regions of the two molecules are complementary in structure."[29] Two years later, in a talk on "Molecular Architecture and the Processes of Life" at the Sir Jesse Boot Foundation in Nottingham,[30] Pauling once again made a plea for the generality of molecular surface complementarity, now directly referring to the possibility of synthetic antibody production. The complementary concept thus may provide "an automatic method of producing a substance with a specific biological property, that of combining with the molecules of the antigen."[31] The medical implications of Pauling's ideas experienced a boost in popularity with his research on the molecular causes of sickle cell anemia.[32] In 1949, a research group under Pauling, including the PhD candidate Harvey Itano, introduced the concept of "molecular diseases," suggesting that certain hereditary blood

---

"Molecular Architecture and the process of life," *21st Sir Jesse Boot Foundation Lecture,* May 28, 1948, Nottingham, England (OSU Special Collections). Pauling, Linus, "Molecular Structure and Intermolecular Forces." In *The Specificity of Serological Reactions,* edited by K. Landsteiner, New York: Dover Publications, 1940, 275-293. For a more detailed analysis of Pauling's generalizing strategy, see Mertens, Rebecca, *The Construction of Analogy-Based Research Programs. The Lock-and-Key Analogy in 20th Century Biochemistry,* Bielefeld: transcript Verlag, 2019, 146-152.

28   Pauling, "Molecular architecture and biological processes," 1376.
29   Ibid.
30   Pauling "Molecular architecture and the process of life," 10.
31   Not only did Pauling use the analogy of antibody-antigen complementarity. He also suggested a very specific mechanism of the general causes of physiological and biological processes, "one of moulding a plastic material, the coiling chain, into a die or mould, the surface of the antigen molecule." According to Pauling, this "same process of moulding of plastic materials into a configuration complementary to that of another molecule, which serves as a template, is responsible for all biological specificity." Pauling, "Molecular Architecture and the process of life," 10.
32   Strasser, Bruno, "Linus Pauling's "Molecular Diseases": between History and Memory," *American Journal of Medical Genetics* 115(2), 2002: 83-93, on p. 88f.

diseases were caused by an anomaly of specific, regulatory macromolecules.[33] They proposed that patients suffering from sickle cell anemia possessed an altered hemoglobin which differed from normal hemoglobin with respect to its electrophysical properties. In analogy to the case of antibody formation, the authors proposed that the electrophysical alteration of sickle cell hemoglobin was caused by a complementary interrelation between the globulin part of the hemoglobin and another surface region of the same molecule. Due to this complementarity, the sickle cell hemoglobin would fold into a different shape than normal hemoglobin, causing the "sickling" of the blood cells.[34] Pauling's attempts to generalize the concept of "molecular diseases" were scientifically controversial but flourishing in post-war science policy and popularization.[35] In the context of research organization at Caltech, however, Pauling's general claims about the molecular basis of biological and medical processes were implemented within a large-scale research program, conducted by research groups in the Biology and Chemistry Division between the early 1940s and the late 1950s. Let us now take a closer look at how this implementation was achieved.

2.   Implementation of projects by means of comparison

From 1941 to 1944, Pauling intensified the collaboration between his research group, working on "chemistry in relation to biological problems,"[36] and several groups in biology, including the geneticists Alfred Sturtevant and Sterling Emerson, as well as the embryologist Albert Tyler. In November 1944, Frank Blair Hanson, former president of the RF and then associative director of the natural science division, encouraged Pauling to explicate the relations of his planned project on proteins and antibodies to the ongoing projects on immunological problems in embryology and serological genetics, conducted by the geneticists Alfred Sturtevant and Sterling Emerson, as well as the em-

---

33   Pauling, Linus, et al., "Sickle cell anemia, a molecular disease," *Science* 110(2865), 1949: 543-548.
34   Pauling, "Sickle cell," 546f.
35   Kay, *Molecular Vision*, 258f.
36   Rockefeller Foundation, *The Rockefeller Foundation Annual Report 1942*, 146.

bryologist Albert Tyler.[37] A couple of months later Pauling responded to Hanson, pointing to the close connections of the previously mentioned research projects with respect to "the great problem of the general structure of proteins."[38] The first draft of an expanded, combined proposal was enclosed in a letter to Weaver in December of 1945, asking for his recommendations prior to the actual application. In this correspondence, the project was characterized as "basic research on the great problems of biology," made possible by "great cooperation with the present activities carried out at the Institute."[39] This coop-project, which was planned for the next two decades, aimed to establish a connection between the fields of organic and biochemistry, immunology, embryology, and genetics and their objects of study. In relation to Pauling's former research on the chemical structure of proteins and his immunochemical work on antibodies, it provided a unified account on problems of biological significance.[40] Crucial for the collaboration on the proposal level was the theoretical and semantic integration of projects involved, such as immunochemistry, enzymology, serological and developmental genetics as well as chemical embryology. The biology department contributed to the larger project with an integrated proposal for a sub-project in "Serological Genetics and Embryology."[41] This project was sketched as one "of wide scope," promising "the possible discovery of concepts of broad biological significance" which should constitute a new field of research, centered around the presence of "complementary substances" active in living organisms.[42] The proposed studies were based on several assumptions concerning the feature of biological processes involved, similar to the ones pointed out by Pauling in the previously mentioned articles on the role of specificity and surface complementarity in biological processes. Claims of similarity between the fields of research involved

37   Letter from Hanson to Pauling, Nov. 24, 1944, OSU Special Collections, Linus Pauling and the Structure of Proteins, accessed May 2020, http://scarc.library.oregonstate.edu/coll/pauling/proteins/corr/sci14.039.2-lp-hanson-19450305-01.html.

38   Letter from Pauling to Hanson, March 5, 1945, OSU Special Collections, accessed May 2020, http://scarc.library.oregonstate.edu/coll/pauling/proteins/corr/sci14.039.2-lp-hanson-19450305.html.

39   Letter from Pauling to Weaver, December 4, 1945, OSU Special Collections, accessed May 2020, http://scarc.library.oregonstate.edu/coll/pauling/proteins/corr/sci14.039.3-lp-weaver-19451204.html.

40   See Kay Molecular Vision, 173.

41   Program Outline on Serological Genetics and Embryology, 1946, Caltech Archives, Biology Division, Box 62, Folder 13.

42   Ibid., "Introduction," 1.

were made on the level of phenomenological similarities in the biological processes under study which, according to the proposal, all "involved reactions in which the components exhibit varying degrees of specificity of interaction."[43] This is especially striking for the depiction of Sterling Emerson's and Alfred Sturtevant's project outline on "genic relationships," induced by mutations in the model organism Neurospora, claiming for a "common structural basis underlying the specificities of genes and antigens."[44] According to Tyler and his group, their research on embryological fertilization in marine eggs involved "primarily the specific interacting substances of eggs and sperms" which were further described as "two complementary substances, one on the surface (the gelatinous coat), the other below the surface of the same cell, that are capable of interaction." [45] They continued that the "presence of two such complementary substances implies that there may very well be more" and that the cell might be "constructed of a mosaic of substances that are pair-wise or multiwise complementary."[46] Common to all outlines of the sub-projects involved was further the tendency to depict research activities with the terminology of "antibody-antigen" interrelations.[47] As I have shown elsewhere in more detail, the understanding of intermolecular interactions as "antibody-antigen-like" was supported by the increasing influence of the "lock-and-key analogy of enzyme-substrate relations" and its application to immunological problems in the first half of the 20th century.[48] The Caltech group, however, explicated the assumptions formerly only vaguely suggested by the analogy. Thus, at the beginning of the proposal, it was already stated that the planned projects in the realm of Serological Genetics and Embryology should be developed from the "general point of view" that "some of the fundamental problems" in this field "are those characterized by reaction specificities analogous to those exhibited in the typical antigen-antibody reaction of immunology."[49] The common feature of the studies involved was thus the characterization of "the underlying reactions ... by a fitting together of the components by complementary surface configurations, as exemplified in the antigen-antibody reaction."[50] These

---

43   Ibid., "Introduction," 2.
44   Ibid., "chapter B.1."
45   Program Outline on Serological Genetics and Embryology, 1946, "chapter A.1."
46   Ibid.
47   Mertens, *The construction*, 158-165.
48   Ibid., Chapters 3 and 4.
49   Program Outline on Serological Genetics and Embryology, 1946, "Introduction," 2.
50   Ibid.

claims of similarity shaped the very organizational structure of the project outline, as shall be once more exemplified by the following excerpt of the list of projects and themes involved:

"Outline of projects

- A. Embryological relationships

    1. Fertilization [...]

    2. Cell Structure and development as related to complementary substances [...]

- B. Genic relationships

    1. Induction of specific mutations by antibodies (see also E.1.) [...]

    2. Induction of specific mutations by substrates [...]

    3. Specific mutation in phage by renaturation on cell surface

- C. Enzymic relationships

    1. of specific enzymes as immediate gene products

- D. Competitive reactions

    1. As an interpretation of dominant and recessive alleles controlling identical reactions

    2. As related to specificities of adaptive reactions

- E. Gene controlled antigens

    1. Erythrocyte antigens and controlled mutation (see B.1.)

    2. Studies resembling Irwin's on dove hybrids

- F. Tissue Specificities

    1. Tissue transplants and tumor development and specificity

    2. Erythroblast transplantation (related to A.1.e.)

- G. Mechanism of antibody formation

    1. Antibody structure as evidenced by univalent antibodies

    2. Search for natural anti-globulin at site of antibody synthesis [...]

- H. Comparative immunology

    1. Antibody formation in invertebrates [...]

    2. Comparison of the complements of different classes of vertebrates [...]

- J. Population serology

  1. Distribution of blood groups [...]

  2. Phylogenetic relationships in natural antibodies

- K Relations to structure and behavior of chromosomes and genes.

  1. Immunological interpretation of synapsis and crossing-over

  2. Opportunities for "physiological cytology."[51]

The integrative approach was successful; in 1948 Caltech received a seven-year grant ($700,000) from the RF in order to expand their program in biochemistry and molecular biology. In the following years, the conceptual framework of surface complementarity and mutually complementary substances affected primarily the research community in immunology and embryology. In the domain of genetics and molecular biology, however, the concept was soon challenged by Watson's and Crick's work on nucleic acids and the increasing importance of the concept of genetic information and sequence specificity.[52] Yet, even despite the fact that a large part of the proposed similarities between the objects of research could not be proven throughout the 1940s, the cooporative program between the two departments and the established connections between the fields of organic and physical chemistry, immunology, embryology, and serological genetics continued to play an important role in the organization of research at Caltech.[53] In the early 1950s, a new interdisciplinary field "Chemical Biology," including "immunochemistry, molecular structure studies, animal virology, bacterial virology, immunogenetics, animal biochemistry and neurophysiology," was institutionalized at Caltech.[54] In the vein of Vannevar Bush's popular OSRD report "Science, the endless frontier,"[55] Caltech's

---

51   Program Outline on Serological Genetics and Embryology, 1946, 2.

52   Strasser, Bruno, "A World in one Dimension. Linus Pauling, Francis Crick and the Central Dogma of Molecular Biology," *History and Philosophy of the Life Science* 28(4) , 2006: 491-512, here 503ff.

53   Kay, *Molecular Vision*, 238f.

54   Letter from L.A. Du Bridge to Ernest Allen, then head of the Division of Research Grants of the National Institutes of Health (Public Health Service), January 15, 1953, Caltech Archives, Biology Division, Box 22, Folder 19. In this letter, Du Bridge, Caltech's President from 1946 to 1969, informs Allen about the funds needed for new "medical research facilities" for the construction "of a new laboratory of chemical biology" at Caltech.

55   Bush, Vannevar, "Science—The Endless Frontier: A Report to the President on a Program for Postwar Scientific Research," *Washington: National Science Foundation*, 1945.

newly established program in chemical biology was described as "basic research" on macromolecular structures, promising systematic applications for medical problems in the near future.[56] By that time, the idea that all kinds of phenomena in the living world could be linked by similarities on the macromolecular level was dominant and powerful in the American science policy discourse.[57] Caltech's leadership took this idea to the next level, using Pauling's research on the molecular mechanism of sickle cell anemia to promote the importance of research on macromolecular complementarity for medical and pharmacological innovations.[58] In 1951, George W. Gray, the established science journalist and staff member of the RF, once more drew the attention to Pauling's research on sickle cell anemia, pointing to its "far-reaching implications" for innovations in the context of medical treatment and to the "remarkably clear evidence ... that life is basically an affair of molecules."[59] Strasser explains the enduring success and visibility of Pauling's research on sickle cell anemia until the present day by the development of different discovery narratives, contextualizing the concept of molecular diseases for diverse scientific and public audiences.[60] Throughout the second half of the 20th century, Pauling's discovery was thus reconstructed by a plethora of stories which conveyed different narratives of why it was far-reaching in the first place, pointing to various conditions for successful medical research, i.e., the

---

For a detailed study on the role of Bush's report in the development of science policy in the US, see Strokes, Donald E., *Pasteur's Quadrant. Basic Science and Technological Innovation*, Washington D.C: Brookings, 1997. Schauz and Kaldewey recently pointed to the integrative role of the concept of "basic research" in the post-war American science and science policy discourse. Accordingly, the term blurred the former dichotomy of "pure" and "applied" science, allowing scientists to emphasize the fundamental nature and unpredictability of their research, while at the same time praising its importance for social and especially medical progress. See Schauz, Desiree, and David Kaldewey, "Transforming Pure Science into Basic Research: The language of science policy in the US." In *Basic and Applied Research. The language of science policy in the 20th century*, edited by D. Schauz and D. Kaldewey, New York: Berghahn Books 2018, 104-142, on p. 123.

56   Lee Du Bridge, Caltech's president from 1946 to 1969, announced the new program and the construction plans for the Norman W. Church Laboratory of Chemical Biology in January 1953. See the News Bureau of the California Institute of Technology, press release, January 9, 1953, Caltech Archives, Biology Division, Box 22, Folder 19.

57   Abir-Am, Pnina, "The Politics of Macromolecules: Molecular Biologists, Biochemists, and Rhetoric," *Osiris* 7(2), 1992, 164-191.

58   Kay, *Molecular Vision*, 238f.

59   Gray, George W., "Sickle Cell Anemia," *Scientific American* 185(2), 1951, 56-59, on p. 59.

60   Strasser, Bruno, "Linus Pauling's 'molecular diseases,'" 84.

financial support of clinical and interdisciplinary research, the "importance of laboratory research" for medical progress, and the "molecular approach to disease therapy."[61]

3. Project organization narratives

As previously shown, comparing different biological processes and phenomena was a crucial tool in substantiating possible relations between RF funded projects at Caltech and their objects of research at the proposal level. These comparisons, however, had to be convincing for the board of trustees involved in the RF's grant decision processes. In order to put Caltech's projects in biology and chemistry in the right position for extended support, the leading scientists of the projects and especially the chairmen of the Biology and Chemistry Division had to maintain a continuous dialogue with the RF science officers, with whom they developed a successful grant management strategy.[62] Both the RF's officers and the scientists involved had to create a public image of the respective projects, one that was compatible with current trends of the Foundation and, most importantly, justifiable to the board of trustees.[63] As will be shown, especially the RF annual reports provided a platform for the Division officers to justify those projects that already had a history of RF funding and to set the ground work for future support. These reports began with a review of the Foundation's president which could focus on scientific breakthroughs and major investments, or on present social and political contexts. The main part of the reports was devoted to the description of projects and fields of research supported under the respective divisions. At the end of World War II, this structure provided a specific narrative context for the depiction and design of post-war science from the Foundation's perspective. In 1946, the leading narrative of the President's Review was one of "lost opportunities" and, most importantly, lost research goals. Imagining what could have been achieved in biology and medicine if the war had not happened,[64] the review served as an opener for a post-war program, incidentally mentioning those areas of research which were at the core of the following outline in

---

61   Ibid., 85f.
62   Kohler, *Partners in Science*, 2.
63   Ibid., 219.
64   Rockefeller Foundation: *The Rockefeller Foundation Annual Report*, 1946, 23f.

"experimental biology": lost opportunities therewith became future opportunities.[65] Another crucial narrative element was the opposition of war-science and "basic science," according to which war-related research lacked scientific value with respect to epistemic novelty, relevance and openness of research outcomes.[66] The review ended with a plea for unity and collaboration across cultural and scientific boundaries. "The challenge of the future" was thus "to make this world one world—a world truly free to engage in common and constructive intellectual efforts that will serve the welfare of mankind everywhere."[67] As for the specific descriptions of supported projects at Caltech, the 1946 report conveyed a story about scientific collaboration and similarities at the level of immunochemical, genetic, and embryological phenomena.[68] From 1941 to 1945, RF funded projects at Caltech were supported with smaller different grants—there were, of course, connections between them, resulting from the RF's program in experimental biology, the institutional affiliation at Caltech and their context of application, as most of the scientists involved carried out "war-related" research in immunology or serology.[69] However, until the mid-1940s, these projects were described as separate ones in the annual reports of the RF, conducted either under the supervision of Pauling in the Gates and Crellin Laboratories of Chemistry or under Sturtevant in the Kerckhoff Laboratories of Biology.[70] This changed in 1946, when the projects supported at Caltech were sketched as a "program of combined research in the fields of biology and chemistry," dedicated to "such subjects as immunochemistry, serological genetics, chemical genetics and X-ray structural chemistry."[71] In

---

65   Ibid.
66   More specifically, war-related research was described as "drawing on the reserves of the past," "using up the supply of basic discoveries which an earlier generation has given them," and "digging recklessly into the stock pile of existing knowledge." (*Rockefeller Foundation Annual Report*, 1946, 23f.)
67   Ibid., 8.
68   As Kay points out, scientific collaboration has been an important goal of the RF's program in the natural sciences since the early 1930s and has been continuously promoted since then. See Kay, *Molecular Vision*, 7.
69   Kay, *Molecular Vision*, 165.
70   Rockefeller Foundation: *The Rockefeller Foundation Annual Report*, 1945, 146f.
71   The respective passage goes as follows: "Work in immunochemistry has developed in close relation to researches in serological genetics carried on by Professors A. H. Sturtevant, Albert Tyler and Sterling Emerson. Professor George V. Beadle has recently been appointed head of the Division of Biology, where such studies are under way. Most biological processes involve reactions in which the components exhibit remarkable specificity, the best known being the antigen-antibody and enzyme-substrate interactions.

the following two years, descriptions of the "combined" or "joint" program at Caltech became a crucial part of the RF's post-war program. The 1947 report dedicated almost three pages to the Caltech group, quoting several extended passages of Pauling's popularizing talk on "Molecular Architecture and Biological Reactions."[72] The project description was full of generalizing claims about the mechanism of surface complementarity, using Pauling's theory of the physico-chemical causes of antibody formation as a point of reference for analogous claims concerning the causes of enzymatic, genetic, bacteriological, embryological processes and drug action.[73] The common theoretical basis and the possibility that it might reflect universal biological principles justified a very broad and expensive long-term analysis of macromolecules involved in various biological processes. Another aspect of the 1947 report worth noticing was the emphasis on the program's previous history and on the continuity of RF support since 1940.[74] The reconstruction of previous grants in terms of

---

Those fundamental problems of genetics and embryology which are characterized by analogous reaction specificities are being studied under their program in serological genetics." (*Rockefeller Foundation Annual Report*, 1946, 134.).

72   Pauling, Linus, "Molecular Architecture and Biological Reactions," *Biological Science* 24(10), 1946, 1375-1377.

73   In the 1947 report, the research process of the Caltech group was reconstructed as follows: "Experimental work during the past few years has tended to substantiate the findings of earlier investigators that the specific biological forces between antibodies and antigens result from complementariness in structure, or the nearly exact fit between the surface configurations of the antibody molecule and the antigen. [...]. It has been possible to measure the closeness of the surface atoms of antigen and antibody by several different methods, all of which show that the two surfaces are in contact to within about one one-hundred millionth of a centimeter. Many other physiological processes are similarly specific, and it seems likely that their specificity can be given similar explanations. The action of enzymes, drugs and bactericidal substances, even the highly specific power of self-reproduction shown by genes, probably have their origin in forces like those which bring about specificity in serological systems. Studies in chemical embryology reveal that the processes of fertilization are very largely analogous to those encountered in the field of immunology. For example, the engulfment of sperm by the egg resembles the phagocytic processes studied by immunologists, and specific substances obtained from eggs and sperm interact in the manner of antigen and antibody. Complementariness in surface configuration of molecules is no doubt involved in the activities of the thousands of genes that carry to us our inheritance from our ancestors." (*Rockefeller Foundation Annual Report*, 1947, 142f.)

74   Rockefeller Foundation: *The Rockefeller Foundation Annual Report*, 1947, 142.

long-term support played an even bigger role in 1948,[75] when Caltech's "joint chemistry-biology program" was provided with a $ 700,000 grant for the next seven years.[76] On the level of research organization and development, the report descriptions increasingly emphasized the integrative nature of the program between 1946 and 1948. For instance, in the president's annual review of 1948, Caltech's program was characterized in terms of "interrelated efforts of physical and biological scientists to gain more precise information regarding the nature and behavior of living matter," realized by "two closely integrated groups from the Division of Biology and the Division of Chemistry and Chemical Engineering."[77] The topic of research collaboration between the groups in biology and chemistry already played an important role in the 1944 report on RF supported projects in immunology.[78] By 1946, however, cooperation was repeatedly framed in terms of integration on the level of personnel, shared goals and research problems.[79] In 1948, the president's review on "Problems of Modern Society" even dedicated an entire chapter to the topic of "purposeful and conscious collaboration," distinguishing different national systems by their collaborative techniques and values.[80] A narrative was born according

---

75   Here, the RF re-called that it has "long given financial aid for these and other natural science projects at the California Institute of Technology." (*Rockefeller Foundation Annual Report*, 1948, 45.).

76   Ibid.

77   Ibid., 47.

78   In this context, Pauling's and Sturtevant's approaches were described as follows: "The groups headed by these two men are working cooperatively on different aspects of immunology. Each is attacking the subject on a broad and somewhat standard front of research, from which important results are practically sure to come." (*Rockefeller Foundation Annual Report*, 1944, 166.)

79   Ibid., 134f.

80   Accordingly, systematic or "purposeful cooperation" depended "upon three broad essentials: 1) Knowledge based on the experience of effective collaboration, involving techniques ranging from simple group effort, business partnerships, corporate organization and community associations to local, regional, national and international political systems, all interrelated and interdependent, 2) An attitude of tolerance not merely of opinion but also of diverse positions and interests which call for moderation in competitive and combative efforts, 3) The will to cooperate, which implies an acceptance of fundamental values overriding personal and group interests or the exigencies of the moment" (*The Rockefeller Foundation Annual Report*, 1948, 21f.). The emphasis on cooperation was not new for the RF; on the contrary, concepts of cooperation were developed in the context of the evolving "political and economic ideology" after World War I (See Kay, *Molecular Vision*, 7). However, after World War II and with the increasing role of

to which systematic integration and communication at the level of research organization leads to the discovery of fundamental principles in the life sciences. This narrative gained importance and was made visible via success stories, told in Caltech's press releases and in popularizing journal articles. In 1948, after Pauling and his co-workers received the seven-year grant from the RF, George W. Gray published a story about the Caltech group in *Scientific American*, focusing on the encounter of the two leading scientists, Pauling and George W. Beadle, the latter of whom became chairman of the biology division in 1946, and their respective strategies of research organization and collaboration.[81] The article focused on the scientific biographies of both men, the "biologically minded chemist" and the "chemically-minded geneticist," leading over to their "joint program of research on the fundamental problems of biology and medicine."[82] In the same year, *Scientific American* published Weaver's essay on "Science and Complexity" which demarcated the complex problems of 20th century science from those of the 18th and 19th century.[83] The paper further distinguished between problems of organized and disorganized complexity, locating the former in biology, medicine and the social sciences and the latter in the realm of 20th century physics and those branches that heavily relied on statistical methods. According to Weaver, problems of organized complexity posed the main challenge for modern science and society. Dealing with them required novel methods of scientific inquiry and work organization, the most important being "new types of electronic computing devices" and the "mixed-team-approach,"[84] both of which were presented as spin-offs

---

the RF as a player in national and international science policy, the concept of cooperation was linked to the idea of "basic science" and the flourishing "linear model of innovation." (See Kaldewey, David, *Wahrheit und Nützlichkeit*, Bielefeld: transcript Verlag, 2013, 360-366.)

81   Gray, George W., "Pauling and Beadle," *Scientific American* 180(5), 1949, 16-21.

82   Gray, "Pauling and Beadle," 19.

83   Weaver, Warren, "Science and Complexity," *Scientific American* 36(4), 1948, 536-545. The essay was re-printed ten years later in the RF annual report 1958 as part of Weaver's review "A Quarter Century in the Natural Sciences." (*Rockefeller Foundation* Annual Report, 1958, 7-15.)

84   In 1949, Weaver and the engineer Claude E. Shannon published "The Mathematical Theory of Communications" at the University of Illinois Press, exploring the semantic implications of the cybernetic concept of communication, information and coding and their practical value for the sciences. (For a detailed analysis of this theory, see Kay, Lily E., "Who wrote the book of life? Information and the Transformation of Molecular Biology, 1945-55," *Science in Context* 8(4), 1995, 609-634, on p. 622f.)

of the cooperation between scientists and the military during World War II.[85] In the whole, the paper conveyed the message that the complex organization of phenomena in the world could be tackled by a certain organization of research practice—scientific problems were thus re-formulated as "problems of strategy" that could ideally be solved by teams or "units" with diverse scientific backgrounds.[86] Texts like the ones Weaver and Gray published in *Scientific American* supported the persuasion of what can be called project organization narratives, linking a certain way of collaborative research organization to the discovery of complex phenomena and making it seem self-evident and natural.

4.   Conclusion: Comparison and narration in research organization practice

Let us return to the question of why Caltech's program in biology and chemistry was successful, especially on the level of grant application. Certainly, Pauling's "aggressive promotion of immunology as a joint venture" and more generally the search for a flagship of America's scientific advancement in the context of the Cold War were decisive factors for the funding decisions of the RF.[87] The focus on the role of comparison and narration in research organization and specifically grant management practice, however, allows us now to look more closely at the semantic dynamics and power relations involved

---

85   Weaver, "Science and Complexity," 6.

86   Weaver used the example of the so-called "operation analysis groups," initially formed to solve war-related problems, to further explain the appeal of mixed teams: "The attempt to answer such broad problems of tactics, or even broader problems of strategy, was the job during the war of certain groups known as the operations analysis groups. (...) These operations analysis groups were, moreover, what may be called mixed teams. Although mathematicians, physicists, and engineers were essential, the best of the groups also contained physiologists, biochemists, psychologists, and a variety of representatives of other fields of the biochemical and social sciences. Under the pressure of war, these mixed teams pooled their resources and focused all their different insights on the common problems. It was found, in spite of the modern tendencies toward intense scientific specialization, that members of such diverse groups could work together and could form a unit which was much greater than the mere sum of its parts. It was shown that these groups could tackle certain problems of organized complexity, and get useful answers" (Weaver, "*Science and Complexity*," 8).

87   Kay, *Molecular Vision*, 168.

in the respective science policy discourse. From this point of view, one of the crucial factors for the continuous support of the Caltech group was that the comparisons drawn in the course of the project's history in the 1940s *worked* and *made sense* at an organizational level. Put differently: I propose that these comparisons were considered to be meaningful and evident, even though the similarities drawn between genetic, embryological, and immunological phenomena in the application of Pauling's unified "complementariness theory" lacked an empirical basis throughout the 1940s and 50s. Hence, a question that follows concerns the activities and processes that led to the validation and manifestation of comparisons between the respective phenomena. I have suggested that the answer is to be found in the consideration of the relationship between practices of comparison and narration in the context of project organization. More specifically, I argued that comparisons drawn between the domains and objects of research of the bio-chemical program at Caltech were supported by *project organization narratives* conveyed in RF reports and in the realm of science popularization. These narratives created path-dependencies and continuities over time between the projects and people involved. They further established a connection between the integration of multidisciplinary projects at an organizational level and the discovery and explanation of phenomena in the living world.

Seeing, Comparing, Narrating
Making-of the *Middle Ages* in the Early History of Art

Joris Corin Heyder

Introductory remarks

In the Routledge encyclopedia of narrative theory, Werner Wolf points out that "[...] the pictorial medium has problems with narrativity and requires a 'reader' who is much more active in (re-)constructing a narrative than would be necessary in verbal texts."[1] His comment brings up a well-known problem in the discussion on narrativity: images have long been considered incompatible with narrative devices and storytelling.[2] In the same Routledge encyclopedia, Jan Baetens sums up two main arguments for this: First, the ideological argument pursuant to which it is easier to achieve visual literacy than verbal; texts were considered as 'higher art.' Second, a mediological argument according to which one distinguishes between "the fictional character of storytelling vs. the non-fictional character of some sorts of images (photographs)."[3]

---

1   Wolf, Werner, "Pictorial Narrativity." In Routledge encyclopedia of narrative theory, edited by D. Herman, M. Jahn, and M.-L. Ryan, London; New York: Routledge, 2005, 431–435, on p. 435.

2   Baetens, Jan, "Image and Narrative." In Routledge Encyclopedia of Narrative Theory, edited by D. Herman, M. Jahn, and M.-L. Ryan, London: Routledge, 2005, 236–237, on p. 236.

3   Baetens, "Image and Narrative," 236. One of the most astonishing discussions of the (non-)documentary character of photographs has been presented by Roland Barthes in his book Camera lucida. Reflections on Photography, where he tackles—at a certain point—the problem of duration, in both history and photography: "A paradox: the same century invented History and Photography. But History is a memory fabricated according to positive formulas, a pure intellectual discourse which abolishes mythic Time; and the Photograph is a certain but fugitive testimony; so that everything, today, prepares our race for this impotence: to be no longer able to conceive *duration*, affectively or symbolically: the age of the Photograph is also the age of revolutions,

Quite apart from the fact that the second argument is difficult to understand because the image's documentary character has long been disputed, and, of course, many sorts of texts can also be understood as non-fictional,[4] it is obvious in both entries on "Pictorial narrativity" and "Image and Narrative" that their authors describe divergent potentials of narration in texts and images. This approach was also followed by Mieke Bal in her Routledge entry on "Visual Narrativity."[5] In the art historical discourse, it was Max Imdahl (1925–1988) who, prominently, sought to characterize a particular medial quality of images in his conception of *Ikonik*[6] that also critically reflects their narrative potentials and limitations. For Imdahl, the visual experience itself had such a strong evidence ("Anschauungsevidenz") that he denied a lossless translatability from visual representations of an autonomous artwork back to a written narrative.[7] On the contrary, Bal seeks to undermine the distinction maintained between the two media of text and image. She differentiates between, first, appearances of visuality in texts, and, second, narrative aspects of visual

---

contestations, assassinations, explosions, in short, of impatiences, of everything which denies ripening.—And no doubt, the astonishment of *"that-has-been"* will also disappear.", cf. Barthes, Roland, Camera Lucida: Reflections on Photography. Translated by R. Howard. New York: Hill and Wang, 1981, 93–94 [italics in original]. For this passage, cf. also: Bann, Stephen, "History: Myth and Narrative. A Coda for Roland Barthes and Hayden White." In Refiguring Hayden White, edited by F. Ankersmit, E. Domanska, and H. Hellner, Stanford, CA: Stanford University Press, 2009, 144–161, on p. 97–98.

4   A clear division between *res factae* and *res fictae* is already dissolving in Eighteenth-century discourse, as Reinhart Koselleck was able to show, cf. Koselleck, Reinhart, "Terror und Traum. Methodologische Anmerkungen zu Zeiterfahrungen im Dritten Reich." In Vergangene Zukunft: zur Semantik geschichtlicher Zeiten, Suhrkamp-Taschenbuch Wissenschaft. Frankfurt a. M.: Suhrkamp, 2017[1989], 278–299, on p. 278–284. Cf. also: Fulda, Daniel, Wissenschaft aus Kunst: die Entstehung der modernen deutschen Geschichtsschreibung; 1760–1860. European cultures: studies in literature and the arts, 7. Berlin [et al.]: de Gruyter, 1996, 223.

5   Bal, Mieke, "Visual Narrativity." In Routledge Encyclopedia of Narrative Theory, edited by D. Herman, M. Jahn, and M.-L. Ryan, London: Routledge, 2005, 629–633.

6   "Der ikonischen Betrachtungsweise oder eben der Ikonik wird das Bild zugänglich als ein Phänomen, in welchem gegenständliches, wiedererkennendes Sehen und formales, sehendes Sehen sich ineinander vermitteln zur Anschauung einer höheren, die praktische Seherfahrung sowohl einschließenden als auch prinzipiell überbietenden Ordnung und Sinntotalität", cf. Imdahl, Max, Giotto. Arenafresken. Ikonographie, Ikonologie, Ikonik. Munich: Fink, 1980, 92–93.

7   Imdahl, Max, "Ikonik. Bilder und ihre Anschauung." In Was ist ein Bild?, edited by G. Boehm, Munich: Fink, 1994, 300–324, on page 308.

depiction, and thus critically encircles the famous *ut pictura poesis*-paradigm: literally "as is painting so is poetry," respectively "painting is wordless poetry, poetry painting with words."[8] The Simonidian/Horatian aphorism was taken up again and again, but one of the most popular objections derives from Gotthold Ephraim Lessing (1729–1781), who discussed the relation of text and image in his powerful essay on the Laokoon group.[9] He was particularly interested in the different experiences of time and space and believed that images are not able to narrate like a text but rather to catch a "pregnant moment."[10] While Lessing was one of the first who distinguished the arts with respect to their medial differences,[11] Bal's approach emphasizes the (un-)productive cross-over, the intermedia-quality of text-image-relations.[12]

8   A critical overview on the genesis and reception of both, the Simonidian and Horatian dictum: Springrath, Gabriele K., "Das Dictum des Simonides: Der Vergleich von Dichtung und Malerei." Poetica. Zeitschrift für Sprach- und Literaturwissenschaften 3–4, 2004: 243–280.

9   "Die blendende Antithese des griechischen Voltaire, daß die Mahlerey eine stumme Poesie, und die Poesie eine redende Mahlerey sey, stand wohl in keinem Lehrbuche [...]," cf. Lessing, Gotthold Ephraim. Laokoon, oder über die Grenzen der Mahlerey und Poesie. Mit beyläufigen Erläuterungen verschiedener Punkte der alten Kunstgeschichte. Berlin: Christian Friedrich Voß, 1766, 3 [preface].

10   "Kann der Künstler von der immer veränderlichen Natur nie mehr als einen einzigen Augenblick, und der Mahler insbesondere diesen einzigen Augenblick auch nur aus einem einzigen Gesichtspunkte, brauchen; sind aber ihre Werke gemacht, nicht bloß erblickt, sondern betrachtet zu werden, lange und wiederhohlter maassen betrachtet zu werden: so ist es gewiß daß jener einzige Augenblick und einzige Gesichtspunkt dieses einzigen Augenblickes, *nicht fruchtbar genug* gewählet werden kann.", cf. Lessing 1766, 24 [my italics].

11   "Lessings Laokoon ist die erste konsequente Ausarbeitung der Medienästhetik, die in der Natur des Mediums die Natur der Kunst begründet sein läßt. So wird hier, was bisher nur eine technische Voraussetzung zu sein schien, zu einem Moment der Kunst selbst," cf. Stierle, Karlheinz. "Das bequeme Verhältnis. Lessings Laokoon und die Entdeckung des ästhetischen Mediums." In Das Laokoon-Projekt. Pläne einer semiotischen Ästhetik, edited by G. Gebauer, Studien zur allgemeinen und vergleichenden Literaturwissenschaft 25, Stuttgart: Metzler, 1984, 23–58, on p. 38. See. also: Schneider, Sabine. "Die Laokoon-Debatte: Kunstreflexion und Medienkonkurrenz im 18. Jahrhundert." In Handbuch Literatur & Visuelle Kultur, edited by C. Benthien and B. Weingart, Boston: de Gruyter, 2014, 68–85, on p. 72.

12   As it is hardly possible to summarize the rampant, more than two-hundred-year lasting debate on the relationship between text and image, I decided to refer to Bal's position as a *pars pro toto* for a poststructuralist and culturalist perspective that works anew towards a dissolution of the medial boundaries.

Putting all this together, it becomes clear that it can neither be the task of this paper to sum up the different levels of visual and pictorial narrativity nor to generally discuss the possible entanglement of images in narration. Instead, I am interested in a concrete historiographical example. It might bring to the fore why I believe that the narrative potential and/or resistance of images in the discussion on narrativism in history in the course of Hayden White, Frank Ankersmit, and many others is still a blind spot worth being analysed in greater detail. To be more precise: The problem discussed in this paper is a second-order problem, here understood as a distinction between "a sequence of actions or events" and their "discursive presentation or narration."[13] It is assumed that every classification of an era already presents a narrative structure and that—from a historiographical perspective—the above questioned narrative potentials of texts and images can be seen as distinctive, maybe even opposed *stimuli* in the conception of historical periods.

My starting point is the period narrative of the so-called *Middle Ages*. The period in question is still today associated with darkness and decline by a large majority of people; whether in films, novels or computer games, popular culture generally paints the image of an era in dirt, waste, stench, and—strangely enough—authenticity.[14] In the professional art historical discourse, it is still not uncommon to understand the *Medium aevum* quite literally as a period between two peaks—classical Antiquity and its so-called Renaissance. I will not discuss the surprisingly well-established and everlasting narrative as such but the medial process of its conceptualization, particularly in one prominent example of the early history of art. Contrary to history as a discipline, the most important sources for art historians were and still are neither texts nor archival material but visual artifacts. In one longer and two briefer parts of this paper, the guiding question will be whether practices of comparing such artifacts in early 'histories' on Medieval art visually thwarts the narrative, described above, or not. Moreover, I will ask whether comparing images perhaps establishes a hidden, non-verbalized narrative in most of the discussed cases.

---

13   Kemp, Wolfgang, "Narrative." In Critical terms for art history, edited by R. S. Nelson, transl. by D. Britt, Chicago: University of Chicago Press, 1996, 58–69, on p. 67.

14   Simmons, Clare A., "Introduction." In Medievalism and the Quest for the "Real" Middle Ages, edited by C. A. Simmons, New York: Frank Cass & Co. Ltd., 2001, 1–29.

1. Seeing

The abbot Jean-Baptiste Dubos (1670–1742) was one of the first who developed a genuine aesthetic of impact ('Wirkungsästhetik') by bringing forward a semiotic argument, in which he reflects the different levels of perception between text and image:

> Les mots doivent d'abord reveiller les idées dont ils ne sont que des signes arbitraires. Il faut en suite que ces idées s'arrangent dans l'imagination, & qu'elles y forment ces Tableaux qui nous touchent & ces peintures qui nous interessent.[15]

What Dubos presents as the central advantage of a visual representation is something that is already inherent in Lessing's conception of the 'pregnant moment.' In the best case this specific moment draws together past, present, and future in only one crucial scene. Texts, however, have to be read sequentially, and, therefore, it stands to reason that it is easier to grasp, for example, the content of four prints with a sequence from the Passion of Christ than reading a corresponding chapter in the Gospels. But how do we recognize the narrative? One has to look for an organized sequence of events present in different kinds of media such as literature, artworks, music, videos etc.

The single events have to be temporarily structured, and there can be no doubt that a biblical text excerpt with St. John's description of the Passion (fig. 1a) meets this criterion just as well as its visual counterpart: for instance, four engravings by Hendrik Goltzius (1558–1616/17) with the *Carrying of the cross*, the *Crucifixion*, the *Entombment* and the *Resurrection* (fig. 1b) placed next to each other and ordered in a reasonable spatial and temporal sequence of the story. For anyone who is familiar with the biblical Passion, it should be easy to immediately identify the *Crucifixion* scene. It serves as a formal marker that would make it even possible to "read" the other three sequences in a very short span of time of—let's say—in only five seconds. At this point, one may feel reminded of Ludwik Fleck's prominent formula: "To see, one has first to

---

15   Dubos, Jean-Baptiste, Réflexions critiques sur la poésie et sur la peinture. 1 of 2 vol. Paris, 1719, 377. Cf. also: Schneider 2014, 74. The formation of the art in the imagination of the artist, is, however, a concept that had already been developed by Giorgio Vasari (1511–1574), cf. Vasari, Giorgio. Einführung in die Künste der Architektur, Bildhauerei und Malerei. Die künstlerischen Techniken der Renaissance als Medien des disegno, edited by M. Burioni. Transl. by V. Lorini. Berlin: Verlag Klaus Wagenbach, 2006, chap. 15, 98–99.

Figure 1a: The Holy Bible, conteyning the Old Testament, and the New. Imprinted at London: By Robert Barker [...], 1611, The University of Pennsylvania Libraries, Annenberg, Rare Book and Manuscript Library, BS185 1611. L65, Beginning of St. John, Chap. 19.

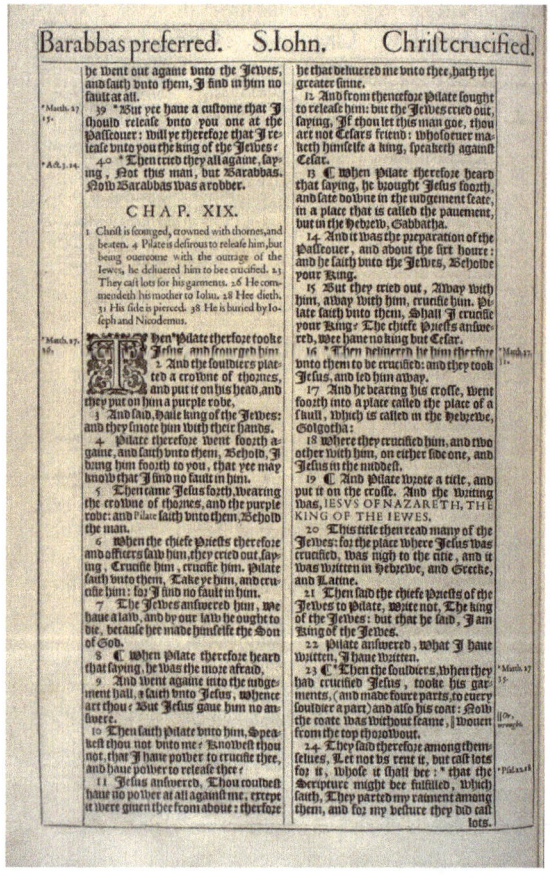

Fig. 1a Schoenberg Center for Electronic Text & Image.

know, and then to know how, and to forget part of the knowledge. One has to acquire a directed readiness to see."[16] It is a thought that can be retrieved again in Bal's understanding of seeing an image through a "cultural frame" which, according to this semiotic approach, creates readability of an image in the first place.[17]

With regard to the here selected beginning of chapter nineteen from the Gospels of St. John in a 1611 print of the King James version (fig. 1a), on the other hand, the experience is different because even if certain narrative structures might be conceivable,[18] we will hardly be able to grasp the entire plot in only five seconds. This is also true because: "reading itself requires constant visualisation."[19] Is it, thus, easier to grasp a narrative in one single moment in a picture series rather than in a text document? While the text becomes blurry if read in a rush, the images keep a certain clarity even in a blink of an eye. However, in images, we do not truly "read" the story but perceive distinct and recognizable forms, or let's say signifiers (for instance, the crucifix or the body of Christ) and perhaps even a multiphase action (crucifix in different positions, as, for instance, on the shoulder of Christ or erected on the hill of Golgotha); both are strong markers for a changing event. Hypothetically, one could object that, for example, a list of catchwords like "Crucifixion," "Be-

---

16   Fleck, Ludwik. Cognition and Fact: Materials on Ludwik Fleck, edited by R. S. Cohen and T. Schnelle, Dordrecht [et al.]: Springer Science & Business Media, 1986, 134.

17   The frame is thought as an "[...] activité sémiotique permanente, sans laquelle aucune vie culturelle ne saurait fonctionner," Bal, Mieke. "Lire l'art?" In Penser l'image. Comment lire les images?, edited by E. Alloa, transl. by M. Boidy, Dijon : Presses du réel, 2017, 43–74, on p. 59. Cf. also : Michalet, Judith. "Sémiotique versus iconique ? Recension à propos de : Emmanuel Alloa (dir.), Penser l'image III — Comment lire les images ?, Les presses du reel," 2017. http://www.laviedesidees.fr/Semiotique-versus-iconique.html [last access : 13.1.2019].

18   This could be, for instance, the use of two proposition following each other, or the perception of a particular time structure within the language, as, for example, the use of past tense in the conjugated verb "tooke" in the first paragraph of chap. 19 in the Gospel of St. John (cf. fig. 1a).

19   Bal, here, is in line with Gérard Genette, cf. Bal 2005, 632. Cf. also: Cooke, Peter, and Nina Lübbren. "Introduction: Narrativity and (French) Painting." In Painting and Narrative in France, from Poussin to Gauguin, edited by P. Cooke and N. Lübbren, London/New York: Ashgate, 2016, 1–21, on p. 9. For Genette's anti-Lessingian conception of the text-based arts not as temporal, but spatial, cf. Genette, Gérard. Narrative Discourse: An Essay in Method. Transl. by J. E. Lewin. Ithaca, NY: Cornell University Press, 1980, 34.

trayal," and so forth might equally be used as markers for a narrative.[20] This is true but only for those readers who are culturally pre-formed; for all others, it makes no difference whether or not the list also contains notions that form no genuine part of the Passion narrative, as, for instance, the "Visitation." On the contrary, for the series of four images, it is potentially possible to realize that there is a connection between the prints with respect to their main characters (Christ), specific objects (cross), and recurring places (Golgotha), even if I do not know what exactly is going on. Furthermore, the number of additional information that are perceivable in images at first glance—whether the *Carrying of the cross* is crowded or not, whether it is day or night, etc.—are not available by looking at a list of catchwords.

In the Western context, the beholder is used to 'read' pictures from left to right and this reliable reading direction supports causal understanding as well. In the chosen example of an album page (fig. 1b) assembled by the famous French connoisseur Michel de Marolles (1600–1681), the 'reading' direction is a rather negligible factor, while, for example, the spatial juxtaposition between the single motifs is much more relevant. The spatial intervals index a temporal sequence and enormously simplify the 'readability.' It does not matter how many intervals have been set between the single represented events to comprehend a commonly known 'story' that has a beginning, a middle, and an end. For my purposes, it is also important to highlight that every interval already stands for a comparison that is performed by the beholder. With Arthur Danto, one could describe events as the raw material of history that produce a difference through time in time; in other words, events create a meaning for an event by relating it to some later or earlier event.[21] It is pos-

---

20   Accordingly, the chapter headings "Barabbas preferred. // S. Iohn. // Christ crucified." in the King James version (fig. 1a) fulfill a comparable task: They give the reader a first orientation in regard to the subject of every page. Typography can be understood as a bridge between text and image that, at least to a certain extent, helps to burst the unison of letters by visually setting accents, gaps, discriminations, and so on.

21   Danto has stressed the analogy between his "model that is representing the structure of a narrative explanation: (1) x is F at t-1, (2) H happens to x at t-2, and (3) x is G at t-3" and causal explanations as such. Moreover, he refers to the Hegelian dialectical pattern: (1) thesis, (2) antithesis, and (3) synthesis, which he understands as well as a narrative structure, cf. Danto, Arthur Coleman, Analytical philosophy of history. Cambridge: University Press, 1968, 233–237. Cf. also: Kemp, Wolfgang, "Ellipsen, Analepsen, Gleichzeitigkeiten. Schwierige Aufgaben für die Bilderzählung." In Der Text des Bildes: Möglichkeiten und Mittel eigenständiger Bilderzählung, edited by W. Kemp, Munich: Edition Text und Kritik, 1989, 62–88, on p. 69–70.

Figure 1b: Hendrik Goltzius, Marolles album, Paris, Bibliothèques nationales, Estampes et photographie, Reserve EC-37-BOITE Fol 1, vol. 1 (of 2), 19.

Fig. 1b Author's photograph.

sible to find the relation between two phenomena and one changing event not only in textual narratives, but also in a picture series and even in a single monophase picture. Narrative techniques such as unfolding, tightening, or tension, brief: the narrative rhythm is able to structure every single picture. It has only to be guaranteed that one character or coherence-creating element like body language, colour, etc. yields causality and chronology.[22]

---

22   Kemp, "Ellipse, Analepsen, Gleichzeitigkeiten," 62.

Figure 2a : Jean-Baptiste Séroux d'Agincourt, Histoire de l'art par les monumens, depuis sa décadence au IVe siècle jusqu'à son renouvellement au XVIe, vol. 5, Paris 1823, plate XI, detail of a female saint.

Fig. 2a https://digi.ub.uni-heidelberg.de/diglit/seroux1823bd6 [last access: 13.2.2019].

But what if an image shows no more than a detail taken out of its original context (fig. 2a)? Does it still have any narrative features? In the chosen example one could say that—after all probability and with a view to the presented veil—we see the head of a woman. Her eyes and mouth are modulated downwards, which may stipulate an emotional reaction in the beholder: this seems to be a rather sad than a happy face, or at least a thoughtful one. The lines in the background could be no more than some ornament, but on the basis of the knowledge of religious Western art these lines form a full circle and thus represent a halo. The description, a text genre which, in my opinion, was erroneously differentiated from narration in the structuralist narratology,[23] maintains narrative potential from the moment when we contextualize it with certain knowledge, in this case the knowledge about veils, halos or

---

23   Saupe, Achim, and Felix Wiedemann, "Narration und Narratologie. Erzähltheorien in der Geschichtswissenschaft, Version: 1.0". Docupedia-Zeitgeschichte. Begriffe, Metho-

Figure 2b : Jean-Baptiste Séroux d'Agincourt, Histoire de l'art par les monumens, depuis sa décadence au IVe siècle jusqu'à son renouvellement au XVIe, vol. 5, Paris 1823, plate XI.

Fig. 2b https://digi.ub.uni-heidelberg.de/diglit/seroux1823bd5 [last access: 13.2.2019].

den und Debatten der zeithistorischen Forschung, 2015. http://docupedia.de/zg/sau pe_wiedemann_narration_v1_de_2015 [last access: 13.2.2019].

Figure 2c : Jean-Baptiste Séroux d'Agincourt, Histoire de l'art par les monumens, depuis sa décadence au IVe siècle jusqu'à son renouvellement au XVIe, *vol. 3, Paris 1823, section "Peintures", 8.*

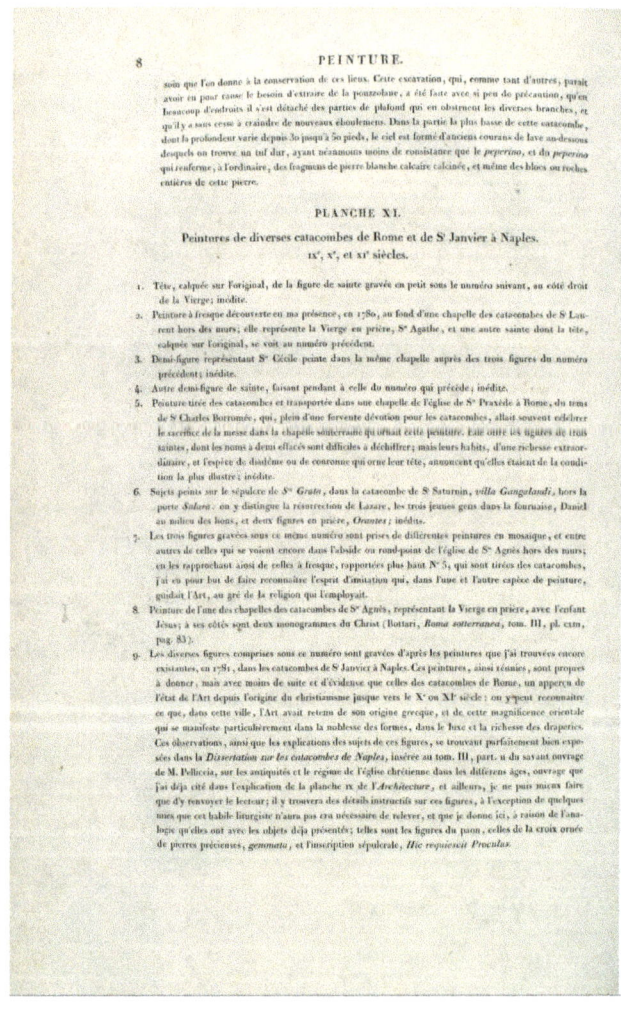

Fig. 2c https://digi.ub.uni-heidelberg.de/diglit/seroux1823bd3 [last access: 13.2.2019].

emotional expressions. Nonetheless, the beholder could only establish some hypothetical narratives, because the detail itself does not show a meaningful change of a situation.[24] The picture's meaning remains obscure as long as it is impossible to connect the detail with a narrative outside itself. As said by Wolf, "a single picture can never actually *represent* a narrative but at best metonymically *point to* a story."[25] According to Danto's equation, this would only be true when the picture cannot offer any indication for a change. If so, and the woman's head (fig. 2a) appears to be a possible example for this, the narrative feature is only achieved by *pointing-at* something (for example, the knowledge about halos). On the contrary, synchronically ordered representations like those in the Six-century Vienna Genesis are unquestionably able to embody a narrative, at least with reference to Danto's analytical pattern of a phenomenon in time (t1) that altered in time (t3) subsequently to a changing event (E) in time (t2).

In one of the miniatures of the Vienna Genesis (fig. 3), Noah is first shown while leaving the Ark together with his family and all the animals, and thereafter only him offering a sacrifice to God. Both scenes are united in an unseparated visual space.[26] In another contribution, Wolf has characterized such cases as (1) multiphase pictures, as distinguished from a group of pictures that, like the four prints by Goltzius described above, forms (2) serial pictures. The detail of the head of a woman (fig. 2a), however, represents in his diction a (3) monophase picture, which—only under certain circumstances—may represent a 'pregnant moment.'[27] His ordering principle is not the only possible way to describe narrative potentials in visual art: already in the late Nineteenth-century, Franz Wickhoff (1853–1909) developed a "general classificatory scheme for visual narration, proposing three principle modes [...]: *komplettierend, kontinuierend and distinguierend* [...]."[28] This comprises representations in which in one scene different events are shown in parallel without depicting figures multiple times—an idea of depiction that can be found on Six-

24   Wolf, "Pictorial Narrativity," 432.

25   Ibid., 433 [italics in the original].

26   Vienna, Austrian National Library, Cod. theol. gr. 31, fol. 2v, cf. http://www.bildar-chivaustria.at/Pages/ImageDetail.aspx?p_iBildID=11470124 [last access: 13.1.2019].

27   Wolf, Werner, "Narrative and Narrativity: A Narratological Reconceptualization and Its Applicability to the Visual Arts." Word & Image 19 (3), 2003: 180–197, on p. 189–192.

28   Cooke, Lübbren, "Introduction: Narrativity and (French) Painting," 4 [italics in the original].

Figure 3: Noah and his family departing the ark and Noah's sacrifice, Vienna, Austrian National Library, Cod. theol. gr. 31, fol. 2v.

Fig. 3 Mazal, Otto. Wiener Genesis: Purpurpergamenthand-schrift aus dem 6. Jahrhundert; vollständiges Faksimile des Codex Theol. Gr. 31 der Österreichischen Nationalbibliothek in Wien mit Kommentarband. Frankfurt a. M.: Insel-Verl., 1980, fol. 2.

century BC black-figure vase paintings and that maybe had its most important renaissance in the futuristic movement of the 1910s. Wickhoff's approach
is so inspiring because with the *complementary*-mode he already implements
the reception process in his model, which has not only been promoted by
Wolfgang Kemp in the late 1980s, but also by Peter Cooke and Nina Lübbren

in their shrewd and recently published paper on "Narrativity and (French) Painting":

> Contrary to Wolf's ideas [...] reception is crucial to constructing narrative meaning [...] if the narrative is in textual format, readers will retell the symbolic letter-signs in verbal form; if the narrative is in pictorial format, viewers will retell iconic signs, also in verbal from [...] the customary opposition of 'word versus image' may be recast as a dialogic circuit in which both words and images operate to similar story-generating ends.[29]

This proposition could still go much further by rather looking for narrative relations instead of asking an object (a text, an artwork) to be narrative. Theoretical framings like Actor-Network-Theory (ANT)[30] or Practice Theory[31] could inspire the discussion on the narrative potentials of text and image beyond their medial differentiation. A narrative network[32] could be one idea that does not weigh between the different *actants*, understood here as human and non-human actors, but which allows us to imagine a *rhizome*-like[33] organized, non-hierarchical interplay of *actants* and their narrative potentials. In such a reductionist/relativist conception, the acting/practicing itself would come to the fore, the seeing or reading, the comparing, and, of course, the narrating. Then, the question would no longer be: 'What are the characteristics of narration?', but, for instance, 'How can we describe the *knowing-how* to retrieve narrative potentials?' and 'Is a narrative potential contingent on processes of routinization?' As for the last question, we could alternatively ask whether or

---

29   Ibid., 10.
30   The relevance of ANT for the analysis of art object-beholder-relations becomes particularly apparent in: Latour, Bruno, Wir sind nie modern gewesen, Frankfurt a. M.: Suhrkamp, 2008.
31   For the intersection of cultural/media theories and practice theory, cf. Reckwitz, Andreas. "Toward a Theory of Social Practices: A Development in Culturalist Theorizing," European Journal of Social Theory 5(2), 2002: 243–263. https://doi.org/10.1177/13684310222225432 [last access: 12.2.2019].
32   Starting from ANT and the adaptive structuration theory (AST), Brian Pentland and Martha Feldman already tried to conceptualize a "narrative network," which they understand as a device for representing patterns of technology-in-use, cf. Pentland, Brian T., and Martha S. Feldman, "Narrative Networks: Patterns of technology and organization," http://citeseerx.ist.psu.edu/viewdoc/download?doi=10.1.1.579.7792&rep=rep1&type=pdf [last access: 12.2.2019].
33   See Deleuze, Gilles, and Félix Guattari, Rhizom. Internationale marxistische Diskussion 67, Berlin: Merve-Verlag, 1977.

not a narrative potential in an image only comes to light when the beholder has already developed a certain 'tacit knowledge' of seeing.[34] This would mean that the act of seeing is rehearsed to complete loose ends or to look for certain narrative markers.

By returning to the detail of the woman's head (fig. 2a), obviously some of its narrative potentials have already been described above. These are forms which represent a halo, a veil, and so on. Yet, their status remains *indeterminate* as long as, say, the beholder has no practice in reckoning such forms. With reference to reception aesthetics, *spaces of indetermination* ("Leerstelle" or "Unbestimmtheitsstelle") have been produced by de-contextualizing the woman's head from its original context.[35] However, the indeterminacy does not just end by re-contextualizing the detail, but, in fact, the rhizomatic structure of potential narratives even multiplies. Seeing the detail of the woman's head in its genuine context (fig. 2b), it is tantamount to *compare* it with a group of other images of much smaller size, which forms part of one and the same visual object. It is neither possible to *ad-hoc* determine the subjects of all the smaller images, nor to recognize a narrative structure at first glance. The relative tininess of the illustrations is one problem, the mode of period style is another, and a third difficulty is how the plate is organized. The individual images appear to reproduce artworks in different sizes and different states of conversation. Flat, linear illustrations are as present as three-dimensional representations and even architectural settings. The plate demonstrates a back and forth between completeness and fragmentation.[36] It forms part of Jean-Baptiste Séroux d'Agincourt's (1730–1814) *Histoire de l'Art par les monumens, depuis sa décadence au IVe siècle jusqu'à son renouvellement au XVIe* posthumously published between 1810–1823.[37] The six-volume book project can be labeled

---

34   For the interrelation between body—particularly the practice of seeing—knowledge, and artifacts, see Prinz, Sophia. Die Praxis des Sehens: über das Zusammenspiel von Körpern, Artefakten und visueller Ordnung. Sozialtheorie. Bielefeld: transcript, 2014.
35   See Kemp, "Ellipse, Analepsen, Gleichzeitigkeiten," 67–79.
36   In the context of the annual congress of the *Nineteenth Century Studies Association (NCSA)* held in Philadelphia in 2018, I sought to conceptualize these notions in my paper "The Connoisseurial Vista. Shifting between Completeness and Fragmentation" as central epistemological practices in 18th-century art connoisseurship, cf. https://www.academia.edu/36237511/The_Connoisseurial_Vista._Shifting_between_Completeness_and_Fragmentation?source=swp_share [last access: 12.7.2019].
37   Séroux d'Agincourt, Jean Baptiste, Histoire de l'art par les monumens, depuis sa décadence au IVe siècle jusqu'à son renouvellement au XVIe [...], 6 vol., Paris : Treuttel et Würtz, 1810–1823. For a meticulously discussion of theses volumes, see Mondini, Da-

as one of the earliest attempts to programmatically shift the interest from a literary to an illustrative art historical approach.[38] It is remarkable, how powerful Séroux d'Agincourt, on the one hand, acknowledges the faculty of seeing and of visual experiences, while, on the other hand, he denies the capability of literature to describe visual phenomena:

> Les productions des Arts fils du dessin, l'Architecture, la Sculpture et la Peinture, consistent en objets sensibles à la vue, sous des formes propres à chacun d'eux, et dont l'effet n'arrive à l'ame que par cet organe ; d'où il résulte qu'on ne doit en écrire ou en étudier l'histoire, qu'en ayant leurs diverses productions sous les yeux [...]. Cependant, parmi les écrivains qui ont essayé de nous faire connaître le sort des Beaux-arts, il en est peu qui aient pris le parti d'en présenter les monumens, et de les laisser parler eux-mêmes aux yeux, en ne les aidant que d'explications succinctes.[39]

Of course, Séroux d'Agincourt, too, does not simply let the images speak for themselves—and it is remarkable how explicitly he suggests that images develop their own language ("laisser parler eux-mêmes aux yeux"). On the contrary, he is not only entitling the plates, like in (fig. 2b): "Paintings from diverse catacombs in Rome and San Gennaro in Naples; 9th, 10th and 11th centuries"[40] and numbering every single image but also trying to contextualize the numbered images in a separate description (fig. 2c). The reference to the women's head (fig. 2a) reads as follows : "Tète, calquée sur l'original, de la figure de sainte gravée en petit sous le numéro suivant, au côté droit de la vierge ; in-

---

niela, Mittelalter im Bild: Séroux d'Agincourt und die Kunsthistoriographie um 1800. Zürcher Schriften zur Kunst-, Architektur- und Kulturgeschichte, Zurich: Zurich Inter- Publishers, 2005.

38   Lena Bader and Johannes Grave outlined the relation between illustration and text in early art historical publications and stressed the fact that "Kunstgeschichte *ad oculus*" was not necessarily understood as a contradiction between image and word, but as something complementary, see Bader, Lena, and Johannes Grave. "Sprechen über Bilder – Sprechen in Bildern: Einleitende Überlegungen." In Sprechen über Bilder – Sprechen in Bildern: Studien zum Wechselverhältnis von Bild und Sprache, edited by L. Bader and J. Grave, Passagen; 46, Berlin: Deutscher Kunstverlag, 2014, 1–30, on p. 8–12.

39   Séroux d'Agincourt, Histoire de l'art par les monumens, 1810–1823, vol. 1, 1.

40   "Peintures de diverses catacombes de Rome et de St. Janvier à Naples. IX.e, X.e, et XI.e Siécles." cf. Séroux d'Agincourt, Histoire de l'art par les monumens, 1810–1823, vol. 5, plate XI (at the bottom of the page).

édite."[41] Apparently, the context of the head-detail now becomes clearer, and even more so the way in which the illustration was made: "calquée" means that the author or someone who accompanied him traced the reproduced figure from the original. In Séroux d'Agincourt's logic, the traced part is shown much bigger than the rest of the images to simulate life-size, and, most importantly, in its facsimileing quality it serves to mirror historical 'factuality.' The same high standard is evident in the second reference that is worth being cited in total:

> Peinture à fresque découverte *en ma présence*, en 1780, au fond d'une chapelle des catacombes de St Laurent hors des murs; elle représente la Vierge en prière, Ste Agathe, et une autre sainte dont la tête, calquée sur l'original, se voit au numéro précédent.[42]

By stressing to have discovered the painting himself, Séroux d'Agincourt expresses not only pride but also his being an eyewitness in a then already 'historicized' past event. The year of the discovery as well as the precise description of the place are two additional factors that show his documentary interest. His efforts appear to stand crosswise to his programmatic intention to narrate a history of art from the "period of its decadence to its renovation," not in the sense that he is doing something else, but by reflecting the negative aftertaste of the project's title. The greatest part of Séroux d'Agincourt's work is dedicated to exactly this kind of supposedly 'lower' Medieval art attentively observed by him and reproduced in a preferably verist, antiquarian manner. Other than in his text corpus, where he follows a Vasarian narrative of decline and renovation, in his illustrations as well as in the paratextual elements, the author seems to forget all his reservations in light of seeing and comparing the originals.

2. Comparing

In a connoisseurial manner, Séroux d'Agincourt not only visited and documented Medieval art but also sought to synthesize art historical relations in plates like the one with two images from the Ghent altarpiece executed by the

---

41   Séroux d'Agincourt, Histoire de l'art par les monumens, 1810–1823, vol. 3, section "Peintures," 8.

42   Ibid., 8 [my italics].

van Eyck brothers and another image that shows a Dead Christ Supported by Angels painted by Antonello da Messina (1429/30–1479) (fig. 4a). By arranging these three images next to each other, he already implements a certain valuating relationship. For example, the numbering of the sequence from one to three could mean a temporal succession or also artistic progress. The caption specifies that the plate's subject is the "invention and practice of oil painting by John of Bruges (i.e. Jan van Eyck) and Antonello da Messina."[43] As a matter of fact, the engraver was not able to particularly illustrate the material quality of oil painting. Without the paratext, however, the meaning of the comparison would have remained puzzling.

Figure 4a : Jean-Baptiste Séroux d'Agincourt, Histoire de l'art par les monumens, depuis sa décadence au IVe siècle jusqu'à son renouvellement au XVIe, *vol. 6, plate CLXXII.*

Fig. 4a https://digi.ub.uni-heidelberg.de/diglit/seroux1823bd6 [last access: 13.2.2019].

What makes things challenging is the fact that it is not the same to analyse 'intended' comparisons in images or texts. The constitutional elements of a comparison are at least two *comparata* which are assumed to be comparable (assumption of comparability or "Gleichartigkeitsannahme") according to

---

43   "Invention et pratique de la Peinture à l'huile, par Jean de Bruges et Antonello de Messine. XV.e Siècle." cf. Séroux d'Agincourt, Histoire de l'art par les monumens, 1810–1823, vol. 6, plate CLXXII (at the head of the page).

Figure 4b : Jean-Baptiste Séroux d'Agincourt, Histoire de l'art par les monumens, depuis sa décadence au IVe siècle jusqu'à son renouvellement au XVIe, *vol. 6, Paris 1823, plate CLXIV.*

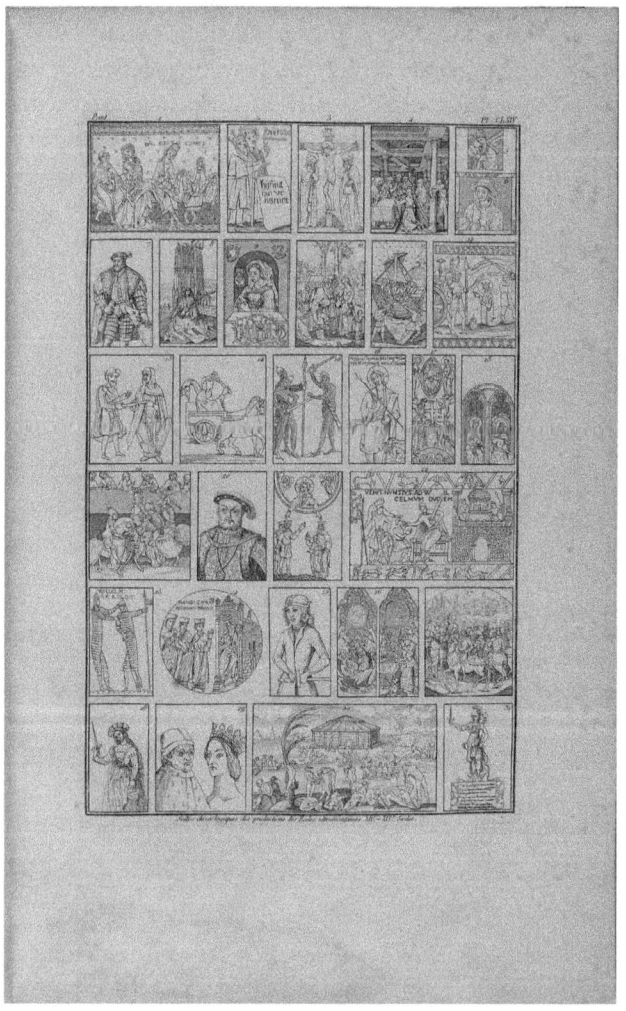

Fig. 4b https://digi.ub.uni-heidelberg.de/diglit/seroux1823bd6 [last access: 13.2.2019].

one or more respects, the *tertium comparationis*. In texts, the *tertium* is often (but, of course, not always) obvious. Images, however, lack the ability to concretize their *tertia*. I could say, for instance, that the reproduction prints one and two of fig. 4a are reproduced details, while number three shows a complete artwork. In this case, the category 'reproduction-print' defines the three *comparata*, while the choice of the picture section of each print is the *tertium*. However, without a guiding text the three images offer a broad range of other possible respects of comparison: Hence, depending on which criterion one focuses on, the comparative arrangement might emphasize drapery or artistic styles as well as thoughts about nakedness in Fifteenth-century art works. It is quite striking to discover the almost endless possibilities of identifying *tertia* in images.

An extreme example for a complex comparison with multiple perspectives is present in another example of Séroux d'Agincourt's *Histoire de l'Art par les monumens*, in which artworks are placed next to each other in stamp-size reproductions (fig. 4b): Neither the topics, the chronological order, the original sizes, nor the direction of reading appear to be the central criterion for the choice and the positioning of the images. The author presents us a colourful tableau of Northern alpine, ultramontane art from the Tenth to the Sixteenth centuries[44] that might express a narrative of artistic progress but fails to do so because of the *surplus* of possible meanings. In the accompanying plate description,[45] the order principle becomes at least clearer, although in an implicit, not an explicit way: The images are presented referring to nations, i.e., it starts with German art and art of the Low Countries, continues with Scandinavian and English art, and finally French art. Once realized, the beholder might be able to compare, for instance, the art of portraiture from England (No. 20: Portrait of King Henry VIII by Hans Holbein the Younger (1497/1498–1543)) with the one from the Southern Netherlands (No. 6: Portrait of Jan de Leeuw by Jan van Eyck (c. 1390–1441)). Even though the temporal distance between these two pictures amount to almost one hundred years, it is not possible to deduce one particular narrative from this comparison, for example, that the later picture is more 'developed' than the ear-

---

44   "Suites chronologiques des productions des Ecoles ultramontaines. XII.e–XIV.e Siècles." cf. Séroux d'Agincourt, Histoire de l'art par les monumens, 1810–1823, vol. 6, plate CLXIV (at the bottom of the page).

45   Every image was described and contextualized by the author, see Séroux d'Agincourt, Histoire de l'art par les monumens, 1810–1823, vol. 3, section "Peintures," 154–158.

lier one. But—pursuant to the choice of n-fold *tertia comparationis*—the doing of comparison helps to rescue narrative potentials. Remarkable is the fact, that—from an analytical perspective—one-dimensional or simple comparisons can be described as follows: the *comparata* (A) and (B) are compared to one another in regard to a single *tertium comparationis* (T) with the result (R).[46] Surely, the typology of comparisons is far more extensive, as Hartmut von Sass and Kirill Postoutenko demonstrated,[47] but this essential structure has an interesting overlapping with Danto's structure of narrative explanations: like Danto's phenomena in time (t1), the *comparata* (A) and (B) in time (t1) face an event in time (t2), i.e., the comparison according to a tertium (T), with the result that the *comparata* have changed to (A)' and (B)' in time (t3). If this analogy is right, then every one-dimensional comparison bears a narrative potential and, therefore, comparative arrangements of visual artifacts offer many starting points for narratives as such, as shown in (fig. 4b).[48]

3.   Narrating

Conversely, in another plate (fig. 5a) Séroux d'Agincourt successfully insinuates a 'Re-naissance' of the 'antiquity' by comparing Raphael's (1483–1520) drawings with some antique fragments. The author offers an imaginative view that lay bare the underlying Vasarian concept of the rise and decline of art.

---

46   I follow here the analytical description by Sass, Hartmut von, "Comparisons. A Typology," 1–15. https://www.academia.edu/37901487/von_Sass_Comparisons._A_Typology [last access: 16.2.2019].

47   See Sass, "Comparisons. A Typology," 1–15, on p. 3, and Postoutenko, Kirill, "Preliminary Typology of Comparative Utterances: A Tree and Some Binaries." In Practices of Comparing. Towards a New Understanding of a Fundamental Human Practice, edited by A. Epple, W. Erhart, and J. Grave, Bielefeld: Bielefeld University Press, 2020, 39–86.

48   I first developed the idea of a structural analogy between practices of comparing and narrating in this paper; however, in the meanwhile, a project group within the framework of the Collaborative Research Centre 1288 "Practices of Comparing. Changing and Ordering the World," Bielefeld University, Germany, has expanded this consideration further. We not only underlined the structural equivalence, but also emphasized the entanglement of 'comparing' and 'narrating' as crucial cultural techniques, see Kramer, Kirsten, Carrier, Martin, Heyder, Joris Corin, and Hochkirchen, Britta, "Vergleichen und Erzählen. Zur Verflechtung zweier Kulturtechniken", Doi: 10.4119/unibi/2942925, Bielefeld 2020 [last access: 29.7.2020].

Figure 5a : Jean-Baptiste Séroux d'Agincourt, Histoire de l'art par les monumens, depuis sa décadence au IVe siècle jusqu'à son renouvellement au XVIe, *vol. 6, Paris 1823, plate CLXXXIII.*

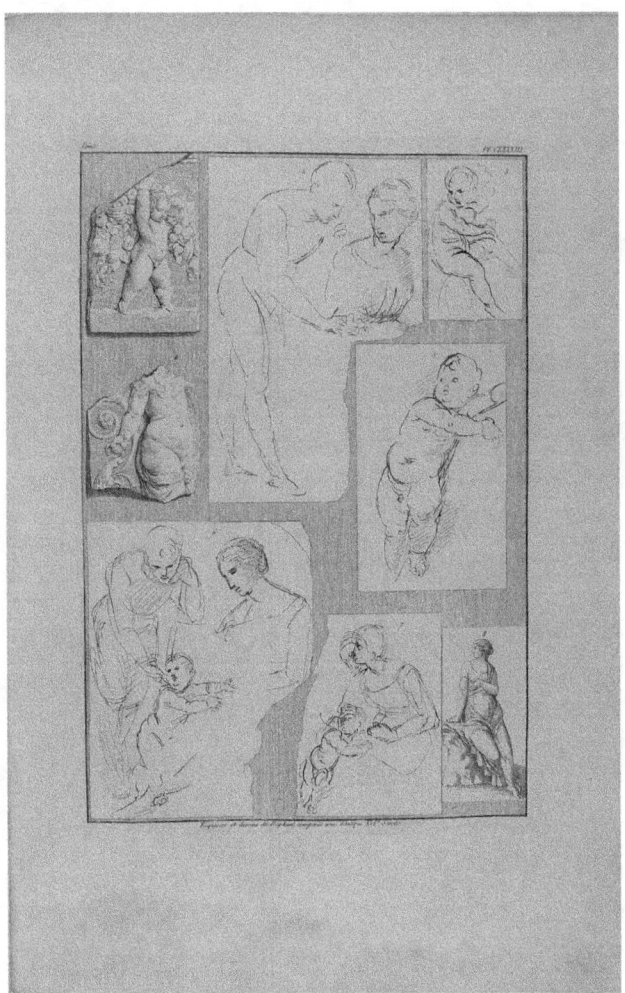

Fig. 5a https://digi.ub.uni-heidelberg.de/diglit/seroux1823bd6 [last access: 13.2.2019].

Figure 5b : Jean-Baptiste Séroux d'Agincourt, Histoire de l'art par les monumens, depuis sa décadence au IVe siècle jusqu'à son renouvellement au XVIe, *vol. 6, Paris 1823, plate CLXXXI.*

Fig. 5b https://digi.ub.uni-heidelberg.de/diglit/seroux1823bd6 [last access: 13.2.2019].

The illustration exemplifies Kemp's idea of "narrative energy" that may particularly arise from a space of indetermination ("Leerstelle"), a gap or a break. In the mentioned example, this gap consists of the long period between the so-called decadence and the renovation. By comparing Raphael's drawings with antique fragments, the narrative energy is unfolded by the implemented historical change that in its result returns to something very similar. Therefore, the plate could be read as an annihilation of the entire book project—that is dedicated to the time in-between (the *Middle Ages*)—, and this is also true for the one that shows a double portrait of Raphael and his teacher Pietro Perugino (1446/1452–1523). The portrait is entitled with the following verse written in capitals: "ENFIN RAPHAEL VINT" (fig. 5b). However, subsequent plates tell a different story that not only stresses the organic entanglement of both Medieval and Renaissance art, but also—for example—a great openness for the affective qualities of medieval artworks. We can perhaps say that every single plate in Séroux d'Agincourt's volumes that is based on comparative dispositions has the potential to disclose its own narrative. This might be a very unsatisfactory result for every approach that is looking for a more general argument towards a pictorial narrativism in historiography. Nevertheless, Séroux d'Agincourt's case shows the weakness of any approach that is only dedicated to texts, as, for instance, White's concept of a meta-narratology,[49] or Ankersmit's metaphorical conceptualization of history as a comparison of:

> [...] one book with another[...]. We do not 'see' the past as it is, as we see a tree, a machine or a landscape as it is. We see the past only through a masquerade of narrative structures (while behind this masquerade there is nothing that has a narrative structure).[50]

I would, instead, propose that images are in parallel able to both offer and to contradict narratives, particularly in cases of non-subsequent picture series. This might also include seeing the past mediated through images, although these images will never be facts, unless in a relativist Wittgensteinian sense, referring to which every fact is *contingent*.[51]

---

49   White, Hayden, Metahistory: The Historical Imagination in Nineteenth-Century Europe, Baltimore: The Johns Hopkins University Press, 1973.

50   Ankersmit, Frank, Narrative Logic. A Semantic Analysis of the Historian's Language, The Hague: Martinus Nijhoff Publishers, 1983, 91–92.

51   Mulligan, Kevin, and Fabrice Correia, "Facts." The Stanford Encyclopedia of Philosophie, 2017. https://plato.stanford.edu/archives/win2017/entries/facts/ [last access:

4. Conclusion

It should be a common practice to discuss the narrative potentials of images in a historiographical approach on early art history, but there have only been sporadic propositions in this direction, for example, by Daniela Bleichmar,[52] Daniela Mondini[53] or Bernd Carqué.[54] Meghan C. Doherty took a major step forward in this respect, by analyzing the individual existence of an illustrative discourse in two late Seventeenth-century journals, the *Philosophical Transactions* and the *Journal des Sçavans*.[55] The main interest, however, is still dedicated to text oriented narratives. Consequently, in a broader historiographical approach it does not come as a surprise that images still play only a minor role in research on the complexity of narrative colligations. Given that not only comparisons in texts but also in images are capable of establishing a narrative, we have to ask in a next step, what kind of implications and limitations are related with comparative viewing? In line with Imdahl, for example, is it possible to say that the life-size proportions and the material quality of the original is undermined by reproduction prints anyway and that it is simply impossible to fully experience the iconic evidence of an artwork on the basis of illustrations?[56]

Later conceptualizations of Medieval art, as, for example, in the famous essay *Das Nachleben der Antike im Mittelalter* by the art historian Anton Springer (1825–1891) from 1867, went into another direction as Séroux d'Agincourt's image-oriented approach. The images were often used with a quite different goal in mind. Springer gave crucial clues by comparing a tiny bronze

---

12.2.2019]. The relation between contingency and narration has already been emphasized by different authors, for instance, see Meuter, Norbert, "Narration in Various Disciplines." In Handbook of Narratology, edited by P. Hühn, J. C. Meister, J. Pier, and W. Schmid, Berlin, Boston: de Gruyter, 2014, 242–262, on p. 257.

52   Bleichmar, Daniela, "Learning to Look: Visual Expertise across Art and Science in Eighteenth-Century France," Eighteenth-Century Studies 46(1), 2012: 85–111. https://doi.org/10.1353/ecs.2012.0084 [last access: 12.2.2019].

53   Mondini, Mittelalter im Bild.

54   Carqué, Bernd, "Epistemische Dinge: zur bildlichen Aneignung mittelalterlicher Artefakte in der Moderne." In Bilder gedeuteter Geschichte, edited by B. Carqué, O. G. Oexle, Á. Petneki, and L. Zygner, Göttinger Gespräche zur Geschichtswissenschaft 23, Göttingen: Wallstein Verlag, 2004: 55–162.

55   Doherty, Meghan C., "Giving Light to Narrative: The Use of Images in Early Modern Learned Journals," Nuncius 30 (3), 2015: 543–569.

56   Imdahl, "Ikonik. Bilder und ihre Anschauung," 320.

Figure 6a: Anton Springer, Das Nachleben der Antike im Mittelalter,
In Bilder aus der neueren Kunstgeschichte, ed. by Anton Springer,
Bonn 1867, 14–15.

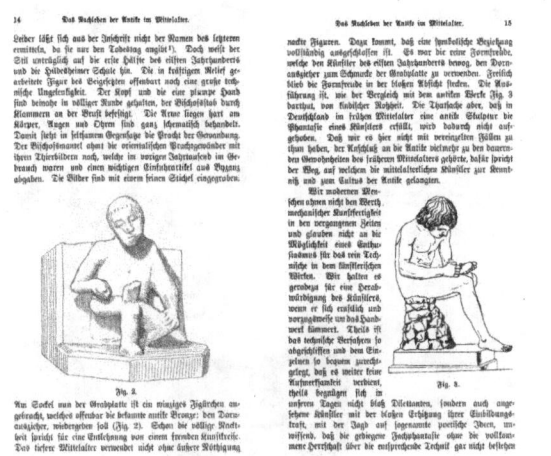

Fig. 6a Springer, Das Nachleben, 14–15.

from a Twelfth century Magdeburgian tomb slab[57] to the antique Capitolinean Spinario (fig. 6a).[58] He followed an understanding of Medieval art according to which antiquity lived on in Medieval art. This new narrative was a veritable shift of paradigm. The comparison is explicitly addressed by the author:

> Die Ausführung ist, wie der *Vergeich* mit dem antiken Werke darthut, von kindischer Rohheit. Die Thatsache aber, daß in Deutschland im frühen Mittelalter eine antike Skulptur die Phantasie eines Künstlers erfüllt, wird dadurch nicht aufgehoben.[59]

---

57 Magdeburg, Dome, Tomb of Frederic I of Wettin, detail of the 'boy with thorn', bronze, c. 1152, cf. http://www.rdklabor.de/wiki/Datei:04-0291-2.jpg#/media/File:04-0291-2.jpg [last access: 12.2.2019].

58 Rome, Conservators's Palace, Sala dei Trionfi, 'boy with thorn', bronze, 5th/1st-century BC, cf. http://www.rdklabor.de/wiki/Datei:04-0291-1.jpg#/media/File:04-0291-1.jpg [last access: 12.2.2019].

59 Springer, Anton, "Das Nachleben der Antike im Mittelalter." In Bilder aus der neueren Kunstgeschichte, Bonn: A. Marcus, 1867, 1–28, on p. 15 [my italics].

Figure 6b: Wilhelm Vöge, "Die Bahnbrecher des Naturstudiums um 1200", In Zeitschrift für bildende Kunst, N.F. 25, H. 8 (1914), 216.

216   DIE BAHNBRECHER DES NATURSTUDIUMS UM 1200

ist anders. In dem Königskopf ist alles gesteigert, äußerlich wirksam gemacht; er ist minder robust, doch soll er lebhafter sprechen (Hochgotik).[1] Die Formen sind kraftloser, aber bewegter; abschüssig schließen die Brauen, die Konturen der Wangen. Und wie ein fliegender Mantel ist das Haar. Auf Pathos der Linie ist alles gestellt; die Form ist der Linie preisgegeben,

anderen frühgotischen Gruppen der Reimser Plastik führen von diesen großen Statuen der Reimser Oberwände die Fäden hinunter. In dem Kopf der Evastatue (neben der Nordrose) lebt das Kopfideal und die schematische Formenbehandlung des »Meisters des Remigiusportales« noch. Man vergleiche den Kopf mit den hier abgebildeten Köpfen (Abb. 29 u. 30), die geschlitzten Augen mit spitzen, leicht geschweiften Winkeln, dem bandartigen Oberlid usw. und auch den Schnitt des Mundes. — Deutlicher noch ist die Zugehörigkeit der r. der Südrose stehenden Königsstatue (mit den Rosenglocken) zu der Richtung mehrerer der Engel von den Chorkapellen und (des weiteren) zu der des Visitatiomeisters.

1) Vgl. m. Bemerk. Repert. f. Kunstwiss. a. a. O., S. 24, 26f.

Abb. 30. Reims, Kathedrale, südlichen Querhaus

die Form ist flächiger.[2] Doch die Augen sind tiefer gebettet; ein stärkerer Schattenschlag ist erzielt; das ganze bewegter und starrer zugleich. Der plastische Geist ist verflogen; dekorativ-monumental ist Trumpf. Auch die Rechnung auf Fernwirkung war im Spiel[3].

Dem »Meister des Peter und Paul« eine solche Wandlung zuzumuten, wäre allein auf Grund eines Rechnungsvermerkes erlaubt. Hier lastet Dunkel.

Noch trügerischer aber ist das Reich der Masken. Ich begnüge mich, eine von ihnen zu zeigen (Abb. 40)[4], die vielleicht von des Meisters Hand ist, ein Stück, trotz dekorativer Behandlung erstaunlich in seiner Durchbildung; eine Ruine, wie Pépin Ruine ist. O, daß man sie verderben läßt in Nässe und Wind, die köstlichen Zeugen von Frankreichs Kunstgeschichte!

2) Das in die Fläche Breiten der Haare, ins Breite Kämmen des Barts.
3) Wie Donatello seine Köpfe für die Höhe auf stärkeren Schattenschlag gearbeitet hat, die Augen begann, einzutiefen u. a. m.
4) Irre ich nicht, von der Fassade des Nordtransepts.

Abb. 40. »Peter und Paulusmeister« (?)
Reims, Kathedrale

Fig. 6b Vöge, "Die Bahnbrecher", 216.

With topoi like "childish rawness," Springer certainly applies older patterns of explanation, while in his illustrative apparatus both artworks encounter each other by means of the specific typographical design of the page, thereby entirely ignoring their actual sizes, which were in fact very different. (fig. 6b) There is much to say about the consequences of the development of new reproduction techniques like photography, but I will close this paper with a final example. In 1914, the art historical medievalist Wilhelm Vöge (1868–1952)—one of the first academic professors who had this specialization—published an

article entitled *Die Bahnbrecher des Naturstudiums* um 1200.[60] From a historiographical position, this article is interesting in several respects, but particularly the use of images is worth studying since they are arranged like a
'second order narrative'. The article finishes not only with a textual comparison of two pieces of art but with a rhetorical ellipse:[61] "Noch trügerischer aber
ist das Reich der Masken. Ich begnüge mich, eine von ihnen zu zeigen [...]."[62]
The author entangles the reader in a narrative network, and, in the end, ex-
poses him or her to the veristically portrayed head of an old man, while Vöge
no longer felt the need to accompany this final image with any words.

---

60   Vöge, Wilhelm.,"Die Bahnbrecher des Naturstudiums um 1200," Zeitschrift für bilden-
     de Kunst 49 (N. F. 25), 1914: 193–216. See also: Heyder, Joris Corin. "Same, Similar, Sem-
     blable: Languages of Connoisseurship," Journal of Art Historiography 16, 2017: 1–15.
61   In this case, the ellipse is not understood typographically but with regard to the literary
     content.
62   Vöge, "Die Bahnbrecher des Naturstudiums um 1200," 216.

Narrating Art History
Practices of Comparing in Exhibitions and Written Surveys with regard to *documenta I*

Britta Hochkirchen

1. Narrating and comparing: a conjunction to achieve temporal organization

The crucial role of narration concerning the construction and understanding of history has been the subject matter of many theoretical surveys and case studies.[1] Narration is not supplementary to history but essential to its construction within the historiographic discourse. Research on this decisive role of narration in the history of historiography is most of the time focused on written texts. This is also the case in the historiography of art history which is mostly concerned with written concepts of the history of the art. In this context, the history of modern art was the subject of many volumes of the late 1940s and 1950s: one famous survey is Werner Haftmann's *Malerei im 20. Jahrhundert* (*Painting in the Twentieth Century*), which was published in 1954. At that time Arnold Bode, the chief curator of *documenta*, asked him to be part of the organizational team of the exhibition.[2] Within the Nazi regime the art

---

1   White, Hayden, *Auch Klio dichtet oder die Fiktion des Faktischen. Studien zur Tropologie des historischen Diskurses*, Stuttgart: Klett-Cotta, 1986. Danto, Arthur C., *Narration and Knowledge*, New York: Columbia University Press, 1985.

2   Haftmann was later also in charge of *documenta II* (1959) and *III* (1964). He became the first director of the Neue Nationalgalerie in Berlin in 1967. Cf. for Haftmann and his role concerning *documenta*: Tietenberg, Annette, "Eine imaginäre Documenta oder Der Kunsthistoriker Werner Haftmann als Bildproduzent." In *documenta 1955. Ein wissenschaftliches Lesebuch*, edited by S. Großpietsch and K.-U. Hemken, Kassel: University Press, 2018, 266–275. Hennecke, Desirée, "Werner Haftmann und die documenta: eine Annäherung." In *documenta 1955. Ein wissenschaftliches Lesebuch*, edited by S. Großpietsch and K.-U. Hemken, Kassel: University Press, 2018, 276–278.

historian Werner Haftmann was—this has been investigated lately—involved as member of NSDAP.[3] Hence, his narration of modern art, as told in *Painting in the Twentieth Century* and still known until today, is striking: Haftmann composes a continuous story of modern art as a "European project" that has "survived" the destruction by the National Socialists and even developed after the Second World War. To narrate this story, which is characterized by the continuity and progress of abstract art, Haftmann makes use of practices of comparing as a discursive strategy to (temporally) connect German art with the abstract tendencies of modern art in Europe and to (temporally) disconnect it from figurative, mimetic modes of former periods of art history, on the one side, and from contemporary Socialist Realism in the GDR, on the other. Against this background, this paper focuses on practices of comparing within the historiography of modern art. It will be shown that the history of the abstract, non-figurative artwork becomes a teleological narration of progress by means of comparing. Practices of comparing are—that is one of the central theses of this essay—essential for the temporal structure of the narration.[4] This temporal structure is highly important concerning historiographical narrations. As Lucian Hölscher once put it: "No historical narrative can do without temporal structures and concepts, which predetermine the general outline of the story."[5] Hence, the temporal structure needs a more precise analysis concerning its production.

Very rarely are other media than texts analyzed with regard to their narrative quality. Hence, Mieke Bal has claimed in her essential survey *Narratology*: "*Narratology* is the ensemble of theories of narratives, narrative texts, images, spectacles, events; cultural artefacts that 'tell a story'."[6] Therefore, I

---

3   Cf. Trinks, Stefan, "Braun, abstrakt." In *Frankfurter Allgemeine Zeitung* (02.03.2020), https://www.faz.net/aktuell/feuilleton/kunsthistoriker-documenta-berater-und-lan gjaehrige-direktor-der-berliner-nationalgalerie-werner-haftmann-in-nationalsozi-alismus-verstrickt-16615552.html (last access 03.05.2020). Rauterberg, Hanno, "Hüter des falschen Friedens." In *Die Zeit* (06.02.2020): 54.

4   See for the importance of the temporal structure of the represented story: Fludernik, Monika, *Erzähltheorie. Eine Einführung*, 3rd ed., Darmstadt: *Wissenschaftliche Buchgesellschaft*, 2010, 44.

5   Hölscher, Lucian, "Time Gardens: historical concepts in modern historiography," *History and Theory* 53, 4/2014: 577–591, on p. 577.

6   Bal, Mieke, *Narratology: Introduction to the Theory of Narrative*, 3rd ed., Toronto/Buffalo/London: University of Tronto Press, 2009, 3. Examples for research about a concept of narration which refers also to other media than texts are: Bal, Mieke, "Telling, Showing, Showing Off." In *A Mieke Bal Reader*, edited by M. Bal, Chicago: The University

will—in a next step—analyze how this narration has been dealt with in another medium: the art exhibition. Although it has been often indicated that
art exhibitions construct art history as well as written texts (if not even with
more *evidentia*),[7] there are only few surveys analyzing case studies concerning
the making of the historiographic narrative.[8] With regard to exhibitions in
the Natural History Museum and the Museum of Modern Art in New York,
Mieke Bal argued in her essay *Telling, Showing, Showing Off* that it is the act of
showing within an exhibition that can be compared to the speech act and its
power to produce "truth": "This exposition, both in the broader, general sense
of 'exposing an idea' and in the specific sense of exhibition, points at objects,
and in that gesture makes a statement."[9] On these grounds I will focus on the
parallels and differences between the narration within a text and within an
art exhibition. This is possible with regard to the project of *documenta I* which
was supervised by Arnold Bode. He was supported by Werner Haftmann who

---

of Chicago Press, 2006, 169–208. Koschorke, Albrecht, *Wahrheit und Erfindung. Grundzü-*
ge einer Allgemeinen Erzähltheorie, Frankfurt a. M.: Fischer, 2012. Cf. with special regard
to the narration within exhibitions: Buschmann, Heike, "Geschichten im Raum. Erzähltheorie als Museumsanalyse." In *Museumsanalyse. Methoden und Konturen eines neuen For-*
schungsfeldes, edited by J. Baur, Bielefeld: transcript, 2010, 149–170. See for the aspect
of narration and pictures: Kemp, Wolfgang, "Ellipsen, Analepsen, Gleichzeitigkeiten.
Schwierige Aufgaben für die Bilderzählung." In *Der Text des Bildes. Möglichkeiten und*
Mittel eigenständiger Bilderzählung, edited by W. Kemp, Munich: edition text + kritik,
1989, 62–88.

7   With regard to *documenta I* these constitutive surveys towards the question of art-
historiography shall be mentioned: Grasskamp, Walter, "Modell documenta oder wie
wird Kunstgeschichte gemacht?," *Kunstforum International* 49, 04-05/1982: 15–22. Foster, Hal, "Museum Tales of Twentieth-Century Art," *Studies of Art* 74, 2009: 253–375. Von
Bismarck, Beatrice, "Der Teufel trägt Geschichtlichkeit oder Im Look der Provokation:
When Attitudes become Form – Bern 1969/Venice 2013." In *Kunstgeschichtlichkeit. Histo-*
rizität und Anachronie in der Gegenwartskunst, edited by E. Kernbauer, Munich: Wilhelm
Fink, 2015, 233–248, esp. 234. Locher, Hubert, "Die Kunst des Ausstellens. Anmerkungen zu einem unübersichtlichen Diskurs." In *Kritische Szenografie. Die Kunstausstellung*
im 21. Jahrhundert, edited by K.-U. Hemken, Bielefeld: transcript, 2015, 41–62, esp. 45.

8   See, for example, Hoffmann, Katja, *Ausstellungen als Wissensordnung. Zur Transforma-*
tion des Kunstbegriffs auf der Documenta 11, Bielefeld: transcript, 2013. And recently with
regard to the historiography of the Nazi regime and its connection to modernism:
Tymkiw, Michael, *Nazi Exhibition Design and Modernism*, Minneapolis: University of Min-
nesota Press, 2018.

9   Bal, "Telling, Showing, Showing Off," on p. 171.

was in charge of the art historical basis and of choosing the works of art that were to be presented.[10]

The destroyed Fridericianum was the venue for this exhibition which took place in 1955, ten years after the end of the Second World War and the Nazi regime. The exhibition was conceived to present the current state of art of the twentieth century. The choice of the location and venue in Kassel was a political statement that was in a dilemma with regard to a narration of postwar art history in Germany: On the one hand, the objective was to show that German postwar art could keep up with international, western modernism, despite the degradation of modern, abstract art to "degenerated art" ("Entartete Kunst") by the Nazis.[11] The similarities between European modern art and German postwar art were to be shown to emphasize a continuity of western modernism. On the other hand, the goal was to mark a discontinuity towards the Socialist Realism of the GDR—the border of which was close to Kassel.[12] Both aims are a common place within the research on *documenta I* and the history of this exhibition, which still takes place every five years in Germany. However, there are hardly any analyses of how these narratives were actually brought into the time-space-constellation of the exhibition room. Which curatorial practices mediated this narrative of continuity and discontinuity at the same time?

From the beginning, the *documenta* was understood as a project to show the state of art after the Great War and the Nazi regime in western Germany.[13] It therefore stood in close connection to Haftmann's treatise about painting in the twentieth century, which was published before (volume 1) and during (volume 2) the time of the exhibition's preparation. Haftmann was not only in

---

10   The exhibition took place from July 15 until September 18, 1955. Confer for the first *documenta*: Kimpel, Harald, *Documenta. Mythos und Wirklichkeit*, Cologne: DuMont, 1997, esp. 248–256. Grasskamp, Walter, "documenta – kunst des XX. jahrhunderts. internationale ausstellung im museum fridericianum in Kassel. 15. Juli bis 18. September 1955." In *documenta 1955. Ein wissenschaftliches Lesebuch*, edited by S. Großpietsch and K.-U. Hemken, Kassel: University Press, 2018, 18–25, 19.

11   This is the main thesis of the chapter "Coming to terms with the past." In Harald Kimpel, *documenta. Die Überschau. Fünf Jahrzehnte Weltkunstausstellung in Stichwörtern*, Cologne: DuMont, 2002, 11–26. Winkler, Kurt, "II. documenta 59 – Kunst nach 1945." In *Stationen der Moderne. Die bedeutenden Kunstausstellungen des 20. Jahrhunderts in Deutschland* (exh. cat. Berlin, Berlinische Galerie Museum für moderne Kunst), 3rd ed., edited by K. Winkler, Berlin: Nicolai Publishing & Intelligence GmbH, 1988, 426–473, 427.

12   Cf. Kimpel, *documenta. Die Überschau*, 11–26 .

13   Cf. Grasskamp, „Modell documenta oder wie wird Kunstgeschichte gemacht?".

charge of choosing the works of art that were to be presented at the *documenta I*, he also wrote the introduction to the documenta catalogue. Here the narration of German postwar art as a narration of modernism, which is central to his survey *Painting in the Twentieth Century*, is repeated and emphasized. In the following in a first step, Haftmann's use of practices of comparing concerning the time structure of the textual narration of art history will be analyzed. In a second step, the focus will be on how narration was transferred to the spatial organization of works of art within the exhibition room. Because of the different materiality and mediality of the narration within a text and an exhibition, it is worth looking at what changes can be registered—especially with regard to the practices of comparing and their qualities of structuring time within the narration. The question therefore is how narration and the practices of comparing change with regard to the medium of narration—written texts and exhibition. Is there a connection between narrating and comparing?

2. Narrating and comparing within written art history: Werner Haftmann's *Painting in the Twentieth Century* and his introduction of the *documenta*-catalogue

In 1954, one year before the opening of the *documenta*, Werner Haftmann published his survey *Painting in the Twentieth Century*. This textbook was accompanied in 1955 by a volume of plates.[14] As the title says, the book focuses on painting and on a special time: the twentieth century. With regard to the order of time within Haftmann's textual narration of the history of modern art, he emphasizes in his introduction that the point of view from which the story will be told is the present: It is the contemporary state of the art that he wants to examine. But with that aim in mind, Haftmann describes the problem of defining the so-called present state of the art: "Because of the overlapping of generations and the modes of expression they embody, the so-called 'present' is a phenomenon of great complexty." ("'Gegenwart' wird durch die Überlagerung von Generationen und der von ihnen getragenen Ausdrucksweisen zu

---

14   Haftmann, Werner, *Malerei im 20. Jahrhundert*, Vol. 2, Munich: Prestel-Verlag, 1955. Tietenberg, „Eine imaginäre documenta," 271-272.

einem sehr komplexen Phänomen.")[15] Haftmann points out that the present is
a complex phenomenon because of the multilayering of generations and their
respective stylistic qualities.[16] The art historian Wilhelm Pinder had already
declared in 1926 that there is a multiple temporality within art history when
one considers all the different generations working at the same time.[17] But following Haftmann's argumentation, these different layers of generations com-
plicate an historical understanding of the present.[18] Haftmann's argument is
based on an understanding of art history as a linear progress[19]—but with different lines at the same time. Therefore, he explains in his introduction the
methodological assumption to "narrate" the history of modern, abstract art

---

15   Haftmann, Werner, *Painting in the Twentieth Century*, Vol. 1, London: Lund Humphries,
      1976, 11. Haftmann, Werner, *Malerei im 20. Jahrhundert. Eine Entwicklungsgeschichte*, Vol.
      1, Munich: Prestel-Verlag, 1954, 10.
16   Haftmann refers to Picasso and Matisse who still influence present art. Haftmann,
      Malerei im 20. Jahrhundert, Vol. 1, 10. 50 years later, the historian Reinhart Koselleck
      defined this quality of layers of time as "simultaneity of the non-simultaneous." Kosel-
      leck, Reinhart, "Einleitung." In *Zeitschichten. Studien zur Historik. Mit einem Beitrag von
      Hans-Georg Gadamer*, 4th ed., edited by R. Koselleck, Frankfurt a. M.: Suhrkamp, 2015,
      9–18, 9.
17   Pinder, Wilhelm, *Das Problem der Generation in der Kunstgeschichte Europas*, Berlin: Frank-
      furter Verlags-Anstalt, 1926. Confer for Haftmann's references to Pinder's theory: Fas-
      tert, Sabine, "'Ich habe als europäischer Historiker geschrieben über europäische Male-
      rei'. Werner Haftmanns Prinzipien der Kunstbetrachtung," in: *Kunst—Geschichte—Wahr-
      nehmung. Strukturen und Mechanismen von Wahrnehmungsstrategien*, edited by Stephan
      Albrecht, Michaela Braesel, Sabine Fastert et al., Munich/Berlin: Deutscher Kunstver-
      lag, 2008, 311-326, esp. 311. Moser, Thomas, "'Kunst [ist das], was bedeutende Künstler
      machen'. Zur Differenzierung zwischen Tradition und Innovation in Werner Haftmanns
      Schaffen der 50er und 60er Jahre," *Helikon. A Multidisziplinary Online Journal*, 3 (2014):
      35–53, esp. 37. See for other concepts of temporality within the historiography of art:
      Karlholm, Dan, "Is History to Be Closed, Saved, or Restarted? Considering Efficient Art
      History." In *Time in the History of Art. Temporality, Chronology and Anachrony*, edited by D.
      Karlholm and K. Moxey, New York/London: Routledge, 2018, 13–25.
18   "It is only by adhering to definite methodological principles that it is possible to find
      'historical' trends in so complex a situation." Haftmann, *Painting in the Twentieth Centu-
      ry*, Vol. 1, 11. "In dieser Vielschichtigkeit nun dennoch 'Ge-schichte' zu erkennen, setzt
      einen bestimmten methodischen Ansatz voraus." Haftmann, *Malerei im 20. Jahrhun-
      dert*, Vol. 1, 11.
19   Cf. Hoffmann, *Ausstellungen als Wissensordnungen*, 95. Hemken, Kai-Uwe, "Kuratorische
      Steuerung kultureller Diskurse: documenta 1955." In *documenta 1955. Ein wissenschaft-
      liches Lesebuch*, edited by S. Großpietsch and K.-U. Hemken, Kassel: University Press,
      2018, 127–167, 131.

as a progress: "These fundamental conceptions or basic purposes provide the plan according to which the historian constructs 'history' out of the complex data he finds in reality; only in this way can he describe the process of temporal growth while dealing with simultaneous phenomena [...]." ("Diese Grundeinsätze und Grundentwürfe nun sind die eigentlichen Bausteine, mit denen es dem Geschichtsschreiber gelingt, aus dem Vielschichtigen das 'Geschichte' aufzubauen und eben als Geschichte zur Darstellung zu bringen, die den zeitlichen Wachstumsprozeß innerhalb der Gleichzeitigkeit alles Daseienden zu beschreiben fähig ist [...].")[20] In this context, he later argues similarly in the introduction of the exhibition catalogue of the *documenta*, adding that the perspective towards the historical development is important with regard to the special position of German art:

> "For example, it would not have been difficult to unite the recent painting, which had prevailed in the European countries since the end of the war, in an exhibition in such a way that the attained points of view and the resulting perspectives would have come to light. But this would not have been sufficient for the peculiar German situation. Rather, it required a broader approach, from the perspective of history, so that this fleeting, ever-changing, indeterminable by itself, one-dimensional point of the 'present' again gains breadth, depth, and multidimensionality."

> "[E]s wäre z. B. nicht schwer gewesen, die seit dem Kriegsende sich in den europäischen Ländern mächtig durchsetzende junge Malerei in einer Ausstellung so zu vereinen, daß die erreichten Standpunkte und die daraus sich ergebenden Perspektiven klar zutage getreten wären. Damit wäre aber eben der besonderen deutschen Lage nicht genüge getan gewesen. Diese verlangte vielmehr einen breiteren Ansatz, aus der Geschichte her, damit jener flüchtige, in ständiger Wandlung begriffene, aus sich allein nicht bestimmbare, eindimensionale Punkt 'Gegenwart' wieder Breite, Tiefe, das Vieldimensionale gewinnt."[21]

Haftmann accentuates the development, the "progress of [...] growth" of art history, also with regard to the development of German art, and wants to trace

20 Haftmann, *Painting in the Twentieth Century*, Vol. 1, 11. Haftmann, *Malerei im 20. Jahrhundert*, Vol. 1, 11. Hemken, "Kuratorische Steuerung kultureller Diskurse: documenta 1955," 131.
21 Haftmann, Werner, "Einleitung." In *documenta. Kunst des XX. Jahrhunderts* (exh. cat. Kassel, Museum Fridericianum), 2nd ed., Munich: Prestel-Verlag, 1979, 15–25, 16. Unless otherwise specified, translations are my own, B. H.

it up to the present. Through producing continuity by means of comparing he declares (and legitimates) his concept of abstract art with the help of "natural" growth, grounded on a causal development.[22] Haftmann wants to show the "complex unity of 'the present'" ("Ganzheitlichkeit von 'Gegenwart'").[23] He emphasizes his perspective as a contemporary narrator who tells the story (of the present!) from an actual contemporary point of view by means of looking backwards into the past. To realize this temporal mode of narration, and to demonstrate "the process of temporal growth while dealing with simultaneous phenomena," he has to arrange his narration by means of continuity and discontinuity through practices of comparing. Angelika Epple and Walter Erhart have argued that different strategies of comparing generate—depending on the situative context—the similarity (and continuity) or the difference (and discontinuity) of two or more *relata*.[24] Comparing is therefore not a neutral *modus operandi* but a practice that unfolds its power by neutralizing its outcome: practices of comparing hence are highly performative.[25] They are not

---

22   A similar argumentation was used by Alfred H. Barr for his exhibition *Cubism and Abstract Art* which took place in the Museum of Modern Art, New York, in 1936. His famous chart, which was reproduced on the cover of the exhibition catalogue, shows the development of all the "isms" with the help of arrows, and it also starts with the 1890s. See for this way of narration of "progress" of art: Voss, Julia, "Wer schreibt *Kunstgeschichte*? Kritik, Kunstwissenschaft, Markt und Museum," *Zeitschrift für Kunstgeschichte* 78, 2015: 16–31, 18. For the discussion of Barr's diagram, see also Klonk, Charlotte, *Spaces of Experience. Art Gallery Interiors from 1800 to 2000*, New Haven/London: Yale University Press, 2009, 135–141; Hal Foster also examines continuities and discontinuities within the narration of modernism in the Museum of Modern Art: Foster, Museum Tales of Twentieth-Century art, 354. Lowry, Glenn D., "Abstraction in 1936: Barr's Diagrams." In *Inventing Abstraction 1910–1925. How a Radical Idea changed Modern Art*, edited by L. Dickerman, London: Thames & Hudson Ltd., 2012, 359–363, esp. 361; Hoffmann, *Ausstellungen als Wissensordnungen*, 95; Brennan, Marcia, *Curating Consciousness. Mysticism and the Modern Museum*, London, MA: MIT Press, 2010, 30–57. Mitchell, W. J. T., *Picture Theory: Essays on Verbal and Visual Representation*, Chicago/London, MA: The University of Chicago Press, 1994, 231–235. For the development of Barr's diagram and its relation to other art historiographical models, see Schmidt-Burkhardt, Astrit, "Shaping Modernism. Alfred Barr's genealogy of art," *Word & Image* 16 (4), 2000: 387–400.

23   Haftmann, *Painting in the Twentieth Century*, Vol. 1, 11. Haftmann, *Malerei im 20. Jahrhundert*, Vol. 1, 11.

24   Epple, Angelika and Walter Erhart, "Die Welt beobachten – Praktiken des Vergleichens." In *Die Welt beobachten – Praktiken des Vergleichens*, edited by A. Epple and W. Erhart, Frankfurt/New York: Campus Verlag, 2015, 7–34, on p. 13.

25   Cf. Epple and Erhart, "Die Welt beobachten," 19.

only dependent on the intentions of a (human) agent, in this case Haftmann, but also on the situative context and its medial, i.e., also material, basis.[26] The *tertium comparationis* is also not a neutral but a chosen aspect of two (or more) chosen *relata*. If we look again at Haftmann's order of narration, it is remarkable that he begins his introduction with a comparison:

> "For the profoundly revolutionary developments in painting, which set in about 1890, cannot be viewed apart from modern mankind as a whole, whose situation they illustrate. [...] It [modern painting; author's note] bears witness to the decline of an old conception of reality and the emergence of a new one. The view of the world that is being superseded today is that which was first shaped by the early Florentine masters with their naïve enthusiasm for the concrete reality of the visible world, which they set out to define. It was the foundation of the Renaissance and of the various styles deriving from it down to Tiepolo, and sustained all the idealisations and stylisations evolved over a period of four centuries. This foundation was first breached by the Romantic movement, and the persistent nineteeth-century attempts at restoration were unable to mend the breach."

> "Die revolutionären und radikalen Prozesse, die in der Malerei seit etwa 1890 in Gang gekommen sind, sind eben nicht isoliert vom Ganzen der modernen Menschlichkeit zu sehen, sie haben repräsentativen Wert für sie. [...] Sie berichten in sinnfällig einsehbarer Form vom Untergang eines alten und der Heraufkunft eines neuen Wirklichkeitsbildes. Es handelt sich um die Ablösung des Wirklichkeitsbildes, das die leuchtenden Geister der florentinischen Frühzeit im begeisterten Vertrauen auf die wirkliche Wirklichkeit des Sichtbaren in ihrer archaisch definierenden Weise langsam heraufhoben und das dann bis hin zu Tiepolo die großen, auf jener 'Renaissance' gründenden Stilwelten trug. Dieser gesetzte Wirklichkeitsgrund, der alle Idealismen und Stilisierungen über vier Jahrhunderte auf sich nahm, begann in der Romantik brüchig zu werden."[27]

26 Cf. Epple and Erhart, "Die Welt beobachten," 20.
27 Haftmann, *Painting in the Twentieth Century*, Vol. 1, 10. Haftmann, *Malerei im 20. Jahrhundert*, Vol. 1, 9. It is striking that Alfred H. Barr's famous introduction within the catalogue to the exhibition of *Cubism and Abstract Art* starts with the same comparison of Renaissance art to declare a discontinuity and thus the basis for the continuity of the progress of abstract art: "Sometimes in the history of art it is possible to describe a period or a generation of artists as having been obsessed by a particular problem. The artists of the early fifteenth century for instance were moved by a passion for imitating nature. In the North the Flemings mastered appearances by the meticulous

Haftmann compares painting during the Renaissance with painting during Romanticism as well as with that of the 1890s, a date which is connected to the advent of Impressionism. These *relata* are compared with regard to the aspect (*tertium comparationis*) of style ("Stilwelten") in which the painting refers to "reality" and how this "reality" is understood ("Wirklichkeitsbild"). By emphasizing these aspects, the art of the Renaissance is put in contrast and temporal discontinuity to that of Romanticism, which is itself compared to the art of the 1890s. Haftmann uses this strategy of comparing to construct a narration of art that is perceived as "old"—the stylistic depiction of Renaissance painters and the mode of figurative mimesis—towards art which is perceived as "new"—the stylistic way of depiction brought up by Romanticism and its mode of abstraction. The act of comparing makes it possible to declare discontinuity and continuity at the same time. Hence, in his introductory text of the exhibition catalogue of *documenta I*, Haftmann argues that contemporary art in Germany has to be understood in relation to modern European abstract art. Even if German art and its modes of abstraction have been suppressed by the Nazi regime, according to Haftmann, contemporary German art can pick up on its early attempts at abstraction through the connection to European modern art:

> "Thus, the question regarding the meaning and purpose of a great art exhibition in Germany ten years after the end of totalitarianism, if it had value in general, should under no circumstances ignore the development process of modern art that has lasted for decades. The widespread idea that a handful—depending on the point of view of the desperate or ingenious—individualities so decisively changed the face of art should be replaced by that other, correct one, according to which a general and legitimately developed transformation of consciousness led to, indeed even forced, those changes. Thus, history came into play, the question of continuity, the documentary. And connected to that—the European, because all this was and is a European process down to the last ramifications, down to the very youngest gen-

---

observation of external detail. In Italy the Florentines employed a profounder science to discover the laws of perspective, of foreshortening, anatomy, movement and relief. In the early twentieth century the dominant interest was almost exactly opposite. The pictorial conquest of the external visual world had been completed and refined many times and in different ways during the previous half millennium." Barr, Jr., Alfred H., *Cubism and Abstract Art* (exh. cat. New York, The Museum of Modern Art), New York: The Museum of Modern Art, 1936, on p. 11.

erations. The task was thus: development and European entanglement of modern art. "

"Es durfte also jetzt bei der Frage nach Sinn und Absicht einer, zehn Jahre nach dem Ende des deutschen Totalitarismus und in Deutschland unternommenen großen Kunstschau, sollte sie Wert fürs Allgemeine haben, gerade der seit Jahrzehnten währende Entwicklungsprozeß der modernen Kunst unter keinen Umständen übergangen werden, um die verbreitete geschichtslose Vorstellung, nach der eine Handvoll—je nach dem Blickpunkt desperater oder genialischer—Individualitäten das Gesicht der Kunst so entscheidend veränderte, durch jene andere, richtige zu ersetzen, nach der eine allgemeine und legitim heraufgewachsene Bewußtseinswandlung jene Veränderungen begründete, ja nahezu erzwang. Damit trat die Geschichte ins Spiel, die Frage nach der Kontinuität, das Dokumentarische. Und zusammenhängend damit—das Europäische, denn all' dies war und ist bis in die letzten Verästelungen, bis in die allerjüngsten Generationen hinein ein europäischer Vorgang. Als Aufgabe stellte sich also: Entwicklung und europäische Verflechtung der modernen Kunst."[28]

The progress of modernist art has to be understood as an entangled European project: For Haftmann, the development of modernist art can be presumed only with regard to the compared similarities between European art and their abstract way of depiction. Here again, Haftmann constitutes continuity by means of comparing German contemporary art of the 1950s to European art of the twentieth century. The aspect in which he assumes similarity, the *tertium comparationis*, is again the non-mimetic, abstract way of depiction, which follows, according to Haftmann, from the "new" concept of "reality." He argues that the modernist painting still represents a relation to the "objective world" but not in the *modus operandi* of a mimetic representation:

"Rather, it was precisely an in-depth mode of experiencing the objective world from a special perspective that changed the entire behavior of modern mankind toward the visual world surrounding it. The insight was that the objective world did not exist so unquestionably that the realm of greater knowledge revealed itself only beyond its appearance, that it was already defined by the nature of the beholder and the way of looking at it, often in unexpected ways."

"Vielmehr handelte es sich gerade um eine vertiefte Erlebnisweise der

---

28   Haftmann, „Einleitung," on p. 18.

gegenständlichen Welt aus einer besonderen, die gesamte Verhaltens-
weise des modernen Menschen zur ihn umstehenden Erscheinungswelt
verändernden Erfahrung. Die Einsicht war, daß die gegenständliche Welt
gar nicht so fraglos existiere, daß sich erst hinter ihrer Erscheinung das
Reich größerer Erkenntnis erschlösse, daß sie sich allein schon durch die Art
des Betrachters und die Weise des Betrachtens in oft unerwarteter Weise
definierte."[29]

To organize the narration with regard to this specificity of non-mimetic referentiality towards "reality", Haftmann begins his survey with the emergence
of the Impressionists in the late nineteenth century. The table of contents
consists of thematic chapters which guide the reader from a discontinuity
(the first meta-chapter is entitled "The Turning Point in Art" ("Die Kunst-
wende"[30]) that is defined by the rise of the Impressionism, followed by other
"isms" of the twentieth century, to the final meta-chapter entitled "The Contemporary Scene. Art since 1945" ("Europäische Gegenwart. Die Kunst der
Nachkriegszeit"). The chapters present the individual "isms" of the twentieth century, according to "nationality" or individual artists: "German Impressionism," "Les Fauves," "Kandinsky and the Rise of Abstract Painting," "Neo-
Realism in Germany" or "The Great Style of Pablo Picasso" ("Der deutsche Im-
pressionismus," "Die Fauves," "Kandinsky und die Entstehung der abstrakten
Malerei," "Der Neorealismus in Deutschland" or "Der große Stil Pablo Picas-
sos"). Katja Hoffmann has pointed out that Haftmann uses distinctions in
style to differentiate between epochs, persons, nationalities and "isms."[31]

At the same time, a linear progress is presented: The *tertium comparationis*
with regard to the non-mimetic referentiality towards "reality" is the basis
for this argumentation without explicit comparisons between the different
"isms." The link between the "isms" and their mode of depiction is emphasized
only through words like "also" and "too." In the chapter "The Early Picasso and

---

29   Haftmann, "Einleitung," on p. 18 and 19.
30   The temporal discontinuity is already emphasized in the first sentences of this chapter:
      "As the nineteenth century was drawing to its end, the peoples of Europe were seized
      with a strange spiritual restlessness. By 1890 this had brought about great changes in
      the prevailing life-feeling and its stylistic expression." Haftmann, *Painting in the Twen-
      tieth Century*, Vol. 1, 17. "Eine eigentümliche Unruhe hatte den Geist der europäischen
      Länder erfaßt, als das 19. Jahrhundert zu Ende ging. Um 1890 bereits hatte sie Lebens-
      empfindung und Stilausdruck weitgehend verändert". Haftmann, *Malerei im 20. Jahr-
      hundert*, Vol. 1, 16.
31   Hoffmann, *Ausstellungen als Wissensordnungen*, on p. 76.

Cubism" ("Der frühe Picasso und der Kubismus"), Haftmann argues implicitly for the growing progress of non-representational style: "Cubism is also a striving to fit the representational elements of the picture into the autonomous order of coloured forms, a task that seemed so urgent to the art theorists during that decade" ("Auch der Kubismus ist ein Ergebnis des Angleichens der gegenständlichen Bildinhalte an das selbstständige Ordnungsgefüge der farbigen Formen, das dem bildnerischen Denken des Jahrzehnts so dringlich schien").[32] At the end of each of four meta-chapters of the German edition, however, Haftmann also adds one section entitled "A Backward Glance" ("Der Blick zurück"), in which he refers to the discontinuity by means of a temporal comparison that is marked by the words "no longer":

> "The painter no longer looks outward to the 'motif', but inward to an emotion that strives to manifest itself in the picture. For this reason, the greatest importance is attached to the cultivation of form and colour. Formal invention replaces thematic invention. The picture becomes an independent organism, an architecture of coloured forms in a non-illusionist, non-perspective space that belongs exclusively to the picture."
>
> "Vom Standpunkt des Malers gesprochen, richtet sich jetzt der Blick nicht mehr auf das Draußen und das 'Motiv', er richtet sich mit aller Kraft auf das, was im Bilde zur Erscheinung gelangen will. Aus diesem Grunde wird die Kultivierung der farbigen und formalen Mittel von größter Wichtigkeit. Die formale Erfindung ersetzt die motivische Erfindung. Das Bild ist ein selbstständiger Organismus, eine Architektur aus farbigen Formen in einer nicht illusionistischen Räumlichkeit, die in ihrer Aperspektivik allein dem Bilde zugehört."[33]

Therefore, Haftmann claims that contemporary modern painting takes place exactly on the edge between figuration and abstraction.[34] The asserted discontinuity is the product of a comparison of style with regard to the mode of pictorial referentiality.[35] On the whole, Haftmann's narration is linear in the order of the "isms," which he refers to in his chapters in their "national"

---

32   Haftmann, *Painting in the Twentieth Century*, Vol. 1, 95. Haftmann, *Malerei im 20. Jahrhundert*, Vol. 1, 139.

33   Haftmann, *Painting in the Twentieth Century*, Vol. 1, 145. Haftmann, *Malerei im 20. Jahrhundert*, Vol. 1, 221.

34   Cf. Haftmann, "Einleitung," on p. 21.

35   Hoffmann points out that Haftmann fulfills an analysis that is based on formalistic characteristics of the works of art. From that it would follow that he argues with stylis-

tradition.[36] There are no figures within this first volume of his survey (despite a few portraits of the artists he mentions) and rarely direct comparisons between the different paintings of the different "isms." On the contrary: Although he claims that there is a relation between European modern painting and German art before and after the Nazi regime, he does not explicitly compare them in his first volume. There are chapters about German art ("Expressionism in North Germany," "The Neue Künstlervereinigung of Munich" or "The Bauhaus Painters" / "Der norddeutsche Expressionismus," "Neue Künstlervereinigung München" or "Die Maler vom Bauhaus") as well as chapters on the other "national" "isms": "Italy and the Modern Spirit," "Russian Suprematism and Constructivism" or "The Italian Contribution" ("Italien und der moderne Geist," "Der russische Suprematismus und der Konstruktivismus" or "Der Beitrag Italiens"). However, Haftmann leaves it up to the reader to compare these "isms." Only with regard to the present state of the art, which is the topic of the last meta-chapter entitled "The Contemporary Scene. Art since 1945" ("Europäische Gegenwart. Die Kunst der Nachkriegszeit"), does Haftmann compare works by artists of different nations.

This strategy becomes even more obvious in the second volume of *Painting in the Twentieth Century*, in which Haftmann makes use of figures to underline his arguments visually. Chapters which solely concentrate on the art of one European country ("Italian Painting between the Wars," "German Painting between the Wars" and "French Painting between the Wars" / "Italienische Malerei zwischen den Kriegen," "Deutsche Malerei zwischen den Kriegen" and "Französische Malerei zwischen den Kriegen") are followed by one ("Painting of the Present" / "Malerei der Gegenwart") that focuses on the underlying similarity of contemporary post-war art of the European countries, including German art. While the works of art are compared within the text, the figures are arranged separately: The works of one artist are in most cases arranged on double pages, so that an immediate comparison between paintings by artists from different countries is not initiated directly.[37] Furthermore, the line-up

tic devices and produces a formalistic historiography. Hoffmann, *Ausstellungen als Wissensordnungen*, 76.

36 Cf. Fastert, "'Ich habe als europäischer Historiker geschrieben über europäische Malerei'. Werner Haftmanns Prinzipien der Kunstbetrachtung," 318.

37 There are some exceptions within the German original edition Haftmann, *Malerei im 20. Jahrhundert*, Vol. 2: Double-page with paintings of Leone Minassian and Rolf Nesch, 452 and 453; Werner Heldt and Eduard Bargheer, 458 and 459; Mattia Moreni and Roberto Crippa, 490 and 491; Victor de Vasarely and Max Bill, 498 and 499; Fritz Glarner

of the different European artists makes it possible to compare them to each other but not—at the same time—to former "isms" of art history. To make this clear: One of the contemporary German artists, Fritz Winter, is presented in the chapter "Painting of the Present" ("Malerei der Gegenwart"), followed by figures showing two of his paintings on a double-page, *Große Komposition (Wandlung)* of 1953 and *Komposition Nr. 5* from 1949 (fig. 1).[38]

Figure 1: Double-page with Fritz Winter's »Große Komposition (Wandlung)« of 1953 and »Komposition Nr. 5« from 1949

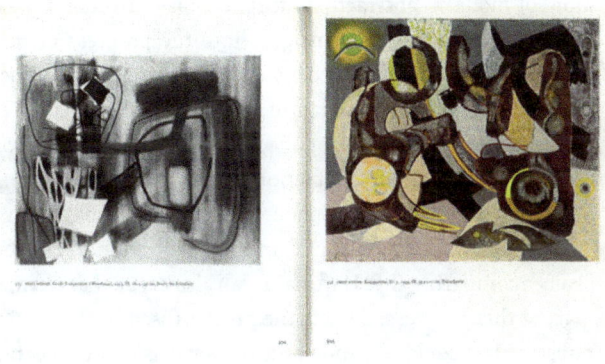

In Werner Haftmann, *Malerei im 20. Jahrhundert*, Vol. 2, Munich: Prestel-Verlag, 1955, 494 and 495.
© VG Bild-Kunst, Bonn 2020.
Creative Commons license terms for re-use do not apply to this picture and further permission may be required from the right holder.

It is striking that the painting on the left is older than that on the right, so that the chronology of "progress" is undermined by the direction of reading. Furthermore, the left figure is black and white while the right one is in color. The comparative perspective does not focus on color but on the abstract mode of depiction. A leading artist of abstraction like Picasso is shown i. a. in the chapter "French Painting between the Wars" ("Französische Malerei zwischen

and Vordemberge-Gildewart, 500 and 501; Giuseppe Capogrossi and Ben Nicholson, 502 and 503.

38   Fritz Winter, *Große Komposition (Wandlung)*, 1953, oil on canvas, 160 x 190 cm, possession of the artist; Fritz Winter, *Komposition Nr. 5*, 1949, oil on canvas, 95 x 110 cm, private collection. Haftmann, *Malerei im 20. Jahrhundert*, Vol. 2, on p. 494 and 495.

den Kriegen")—followed by ten figures on double-pages within the image-part: for example, *Der Hahn* (1938) vis-à-vis *Guernica* (1937) as well as *Nachtfischen bei Antibes* (1939) vis-à-vis *Die Frauen von Algier* (15. Version) (1955).[39] The immediate comparison between the contemporary works of Winter and those of Picasso is not initiated by how the pictures are organized or by the thematic order of chapters. Haftmann's introductory text of the exhibition catalogue of *documenta I*, however, emphasized this comparison with regard to the aspect of similarity in the non-mimetic and non-figurative referentiality.

As a first conclusion of the strategy of narration within texts, we can see that Haftmann uses practices of comparing for the temporal organization of his "narration" of modern, abstract painting. He takes a present perspective but looks back at the past to gain insights into the progress of modernist painting. By means of comparing he constructs a discontinuity of modernist painting that emerged with the Romantic movement and was different than anything before. The difference is proponed with regard to the style of depiction: there is no longer a mimetic depiction of the world (like in Renaissance painting). The comparison between modernist paintings is based on this *tertium comparationis*: They show a continuity of similarity in the way of depiction. And it is through comparing that Haftmann can also argue that German post-war art is part of this progress: despite the break caused by the Nazi regime, modernist abstract art could continue as it is shown as an European project.

The discontinuity Haftmann claims between the mode of referentiality in paintings before and after the Romantic movement, especially with the rise of the Impressionism, is constructed in the narration of the postwar art in parallel to another break. This narration is again structured by means of practices of comparing. In the final chapter ("Present-Day Painting" / "Malerei der Gegenwart") of the second volume, Haftmann compares the present art of western parts of Europe to that of the GDR. The *tertium comparationis* is again the mode of depiction. In this case, Haftmann claims that there is discontinuity and emphasizes this by means of comparing: "Needless to say, they [the modernist painters and the abstract way of depiction; author's note] did not

---

39   Haftmann, *Malerei im 20. Jahrhundert*, Vol. 2, on p. 426–435. Pablo Picasso, *Der Hahn*, 1938, pastell, 56. 5 x 77.5 cm, The Museum of Modern Art, New York; Pablo Picasso, *Guernica*, 1937, oil, 350 x 716 cm, The Museum of Modern Art, New York; Pablo Picasso, *Nachtfischen bei Antibes*, 1939, oil, 206 x 346 cm, The Museum of Modern Art, New York; Pablo Picasso, *Die Frauen von Algier* (15. Version), 1955, oil, 145 x 146 cm, possession of the artist.

entirely eliminate representational painting, but to judge by the general situation of painting today, the capacity of the object to convey ideas and forms has diminished appreciably." ("Selbstverständlich verdrängten sie [the modernist painters and the abstract way of depiction; author's note] nicht die Malerei, die sich auf die sichtbaren Bilder der Wirklichkeit richtete, aber es ist angesichts der allgemeinen Situation der Malerei heute nicht zu leugnen, daß die ideelle und formale Tragkraft des Gegenständlichen ganz fühlbar nachgab").[40] The style of the GDR, the Socialist Realism, with its mimetic referentiality is put in total contrast to that of modernism. Through the comparison, Haftmann creates a discontinuity which connects the style of the Socialist Realism to the past and that of western modernism to the present, if not to the future. To underline this, Haftmann presents an example of Socialist Realism: Renato Guttuso's *Landnahme in Sizilien* from 1949–50 (fig. 2).[41] The painting shows in a figurative style a landscape as well as precisely depicted and differentiated characters in the foreground. Haftmann proposes the comparison in his survey which again emphasizes the progress of abstract art:

"This becomes clear the moment we examine any of the works inspired by Socialist Realism. *The Occupation of Uncultivated Land in Sicily* [...] by the highly-talented and vigorous Sicilian artist Renato Guttuso may serve as an example. The artist is unable to transform his social enthusiasm into a compact pictorial reality, and the result is not a viable realism but hollow declamation. And yet the definition of reality in our century had found a highly-precise expression in Picasso, Beckmann, and Léger. Here was a practicable point of departure, and a good many young artists found their way to this possibility of rendering the experience of reality."

"Das wird unmittelbar anschaulich, wenn man einmal ein Werk des sozialistischen Realismus betrachtet—und als Beispiel stehe hier die 'Landnahme in Sizilien' des hochbegabten und blutvollen Sizilianers Renato Guttuso [...]—das Pathos am Ereignis in seiner lebendigen Wirklichkeit verwandelt sich nicht in bildnerisch verdichtete Wirklichkeit, in einen tragkräftigen Realismus. Nun war es doch aber so, daß die Wirklichkeitsbestimmung in unserem Jahrhundert einen ganz präzisen Ausdruck gefunden hatte bei Picasso,

40   Haftmann, *Painting in the Twentieth Century*, Vol. 2, 439. Haftmann, *Malerei im 20. Jahrhundert*, Vol. 2, 438.
41   Haftmann, *Malerei im 20. Jahrhundert*, Vol. 2, on p. 438. Renato Guttuso, *Landnahme in Sizilien*, 1949–50, ca. 220 x 300 cm, possession of the artist.

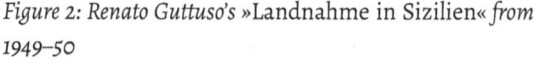

Figure 2: Renato Guttuso's »Landnahme in Sizilien« *from 1949–50*

In Werner Haftmann, *Malerei im 20. Jahrhundert*, Vol. 2, Munich: Prestel-Verlag, 1955, 439. © VG Bild-Kunst, Bonn 2020. Creative Commons license terms for re-use do not apply to this picture and further permission may be required from the right holder.

bei Beckmann, bei Léger. Hier ließ sich wirklich anknüpfen und ein erheblicher Teil der jüngeren Maler fand zu dieser Darstellungsweise von Wirklichkeitserlebnissen."[42]

Haftmann ends his narration of modern, abstract art in the first volume with a chapter that clearly refers to the future, entitled "A Glance Forward" / "Ein Blick nach vorn" (in contrast to the former chapters "The Backward Glance" / "Der Blick zurück"):

---

42   Haftmann, *Painting in the Twentieth Century*, Vol. 2, 439. Haftmann, *Malerei im 20. Jahrhundert*, Vol. 2, 439. In the following, Haftmann refers to an example showing how younger generations connect to the way of depiction proposed by Picasso and the others and not to the one brought forward by Socialist Realism: He shows a figure of Giuseppe Zigaina's *Gras für die Kaninchen*, also from 1949 (like the one by Guttuso). In the original German edition the picture is shown on the next page (440) so that no immediate comparison between the paintings of Guttuso and Zigaina is possible. The reader is forced to turn the page.

"Our path has led us to the present day, but it is not ended. Everything in the art of today points to an unknown future. Post-war painting as a whole has a peculiarly fragmentary character. To be sure, it takes its place within the great stylistic pattern that has been shaping itself since the turn of the century. But within the general pattern it concentrated overwhelmingly on abstraction. Contemporary artists have greatly enlarged and defined the abstract domain."

"Unser Weg ist ausgeschritten, aber nicht zu Ende. Die Grenzmarke des Vorhandenen haben wir erreicht, aber sie schließt nichts ab. Über sie hinaus drängt alles in das noch Ungeborene der Zukunft. Die Gesamterscheinung der europäischen Nachkriegsmalerei hat einen eigentümlichen Fragmentcharakter. Zwar ist sie eingeschrieben in den großen Stilentwurf, der das Ergebnis der bildnerischen Tätigkeit seit der Jahrhundertwende war, aber sie hat innerhalb dieses Stilentwurfes mit Beharrlichkeit das abstrakte Gebiet sich zur Domäne gemacht. Sie hat es erstaunlich erweitert und präzisiert."[43]

As a conclusion, Haftmann shifts his view from the present towards the teleologically framed future. It is the *documenta* exhibition that opens the possibility to experience the upcoming modernist art and its non-mimetic referentiality in the mode of depiction. Haftmann refers to this exhibition in the introduction of the catalogue with an explicit invitation to compare: "Here we now have the first encounter in Germany between younger German art and artists of the other European countries. This is a great and wonderful event for us. We can now compare for the first time how the European countries relate to each other in their contemporary expressions of art." ("Hier vollzieht sich jetzt für uns in Deutschland die erste Begegnung der jüngeren deutschen Kunst mit den Künstlern der anderen europäischen Länder. Das ist für uns ein großes und wundervolles Ereignis. Wir können nun zum ersten Mal vergleichen, wie sich die europäischen Länder in ihren heutigen Kunstäußerungen zueinander verhalten.")[44] These sentences refer to practices of comparing as a mode to continue the narration of the progress of the non-mimetic art of modernism. It is the presence of the exhibition that offers access to this narration.

---

43   Haftmann, *Painting in the Twentieth Century*, Vol. 1, 374. Haftmann, *Malerei im 20. Jahrhundert*, Vol. 1, 478 and 479.

44   Haftmann, "Einleitung," on p. 25.

3.   Narrating and comparing within the art exhibition: *documenta I*

The *documenta I* exhibition of 1955 staged the narration of abstract, non-mimetic art:[45] Bode, on the one hand, kept the raw walls of the destroyed Fridericianum while, on the other hand—and in contrast to these signs of destruction—he exaggerated the exhibition's space by means of textile panels and metal rods on which the artworks were presented—in distance to the disintegrating walls.[46] The art historian Charlotte Klonk has therefore pointed out that this curatorial practice could be understood as indicating a break-up—or in the context of our argumentation: discontinuity—between the presented artworks and the traces of history.[47] Starting from this observation, the *documenta* will be examined in the following with regard to the curatorial strategies with which the narration of postwar art history is produced. In the previous chapter we followed Haftmann's strategies to narrate a teleogical story of progress with regard to the modernist non-mimetic image. By means of practices of comparing, the written narration was structured concerning discontinuity (mimetic art before Romanticism and contemporary art like Socialist Realism) and continuity (non-mimetic art before the Nazi regime and European abstract art). But how was this narration transferred into the exhibition space? In what way did curatorial practices of comparing constitute the temporal organization of the narration concerning continuity and discontinuity?

The *documenta I* presented ca. 150 works of art by European modern artists,[48] which in most cases were also referred to in Haftmann's *Painting*

---

45   Cf. Grasskamp, "Modell documenta oder wie wird Kunstgeschichte gemacht?". At this point, reference is made only by way of example to research literature on *documenta: documenta – Idee und Institution. Tendenzen – Konzepte – Materialien*, edited by M. Schneckenburger, Munich: Bruckmann, 1983. Documenta. Curating the History of the Present, edited by N. Buurman and D. Richter, in *OnCurating* 33 (June 2017).

46   See for a closer examiniation of Bode's display: Hemken, "Kuratorische Steuerung kultureller Diskurse," 136–145.

47   Klonk, Charlotte, "Die phantasmagorische Welt der ersten documenta und ihr Erbe." In *Die Ausstellung. Politik eines Rituals*, edited by D. von Hantelmann and C. Meister, Zurich/Berlin: Diaphanes, 2010, 131–159, 134. See also for the *documenta* and its curatorial practices to produce a spatial experience (mostly with regard to the second edition): Klonk, *Spaces of Experience*, esp. 174-175.

48   See for this data Winkler, "II. documenta 59", 427.

in the Twentieth Century.[49] Additionally, Bode exhibited contemporary artists, again in reference to Haftmann's survey: Wols, Ernst Wilhelm Nay, Fritz Winter, Emilio Vedova or Pierre Soulages.[50] In comparison to Haftmann's written narration, the exhibition deals with the physical presence of the artworks which—brought together in the exhibition space—produce a mixture of different times and temporalities: but within the immediate presence of the exhibition.[51] The linear progression of the text is—at least partly—replaced by the simultaneity of the exhibition. Recent surveys have emphasized the anachronistic character of exhibitions: In an exhibition, different times on different levels—for example, concerning the date of the artwork, the experience of the visitor, the represented time—come together.[52] And in this "simultaneity of the non-simultaneous,"[53] there is no narrator who argues for a special temporal perspective as Haftmann did in the introduction of his survey. The speaker of the exhibition—on the whole, most exhibitions do not have an explicit narrator as Mieke Bal has pointed out[54]—instead places emphasis on the overall presence of the artworks. The performativity of the act of comparing focuses on the quality of evidence, as the artworks—and therefore the narration—are literally (physically) in front of one's eyes. Johannes Grave has argued that practices of comparing also influence the reception of the temporality of pictures and, connected to that—this paper argues—the historiographical attribution of the work of art.[55]

---

49   Therefore also Annette Tietenberg asks how the art historical narration of *Painting in the Twentieth Century* had an impact on the presentation of the artworks within the exhibition (271). She identifies Haftsmann's written survey as "script" for the exhibition (273). Tietenberg, "Eine imaginäre documenta".

50   A main *difference* between the exhibition and Haftmann's survey on painting is the integration of sculptures within *documenta I*.

51   Cf. Pomian, Krzysztof, *Der Ursprung des Museums. Vom Sammeln*, Berlin: Wagenbach, 1986, 44.

52   Cf. Bismarck, "Der Teufel," 234–235. For the question of the special quality of time in exhibitions, see: Bismarck, Beatrice von, Frank, Rike, Meyer-Krahmer, Benjamin, Schafaff, Jörn and Weski, Thomas, *Timing. On the Temporal Dimension of Exhibiting*, Berlin: Sternberg Press, 2014.

53   Koselleck, "Einleitung," 9.

54   Cf. Bal, "Telling, Showing, Showing Off," 174.

55   Cf. Grave, Johannes, "Vergleichen als Praxis. Vorüberlegungen zu einer praxistheoretisch orientierten Untersuchung von Vergleichens." In *Die Welt beobachten – Praktiken des Vergleichens*, edited by A. Epple and W. Erhart, Frankfurt/New York: Campus Verlag, 2015, 135–159, on p. 150-151.

Figure 3: Installation shot of »documenta I«, *1955, Museum Fridericianum, Kassel*

© documenta Archiv (Dauerleihgabe der Stadt Kassel) / Foto: Günther Becker.

With this in mind, it is striking that the exhibition narrates the history of art in the twentieth century in a different way than Haftmann's text.[56] The exhibition is only focused on the temporal dimension of continuity, while the dimension of discontinuity is left out. Paintings which follow a mimetic way of depiction are not integrated into the exhibition: there are no works from before Romanticism as well as no works of Socialist Realism.[57] While Haft-

---

56   Tietenberg emphasizes more the similarities between Haftsmann's written survey (volume one and two) and the exhibition: Tietenberg, "Eine imaginäre documenta," 273. See for a close examination of the exhibition's narration with regard to continuity (87), discontinuity (98) and practices of comparing (113): Grasskamp, Walter, *Die unbewältigte Moderne. Kunst und Öffentlichkeit*, Munich: Beck, 1989, esp. 76–145.

57   Cf. Klonk, "Die phantasmagorische Welt," 142. Furthermore—although following an abstract mode—also works of art by Jewish artists are not exhibited. Cf. Grasskamp, *Die unbewältigte Moderne. Kunst und Öffentlichkeit*, 96.

Figure 4: Installation shot of »documenta I«, *1955, Museum Fridericianum, Kassel*

© documenta Archiv (Dauerleihgabe der Stadt Kassel) / Foto: Günther Becker.

mann narrates the story of modernism by comparing it within his text to the art of Renaissance painting—and visually to the art of Social Realism (with the visual example of Renato Guttuso's *Landnahme in Sizilien*)—in order to highlight differences in modes of depiction and referentiality, the exhibition lacks these comparisons and counter-examples. Only the exhibition space—the destroyed walls of the Fridericianum—conveys a trace towards discontinuity.[58]

In contrast, the *documenta I* exhibition highlights the narration of temporal continuity by means of comparing. While in Haftmann's *Painting in the Twentieth Century* the comparison between works of art before and during the World Wars was not directly initiated because of the linear and separated structure of the chapters of the book (and its visual apparatus), the exhibition emphasizes this comparison with the help of curatorial practices. The famous example is the presentation of Pablo Picasso's *Girl before a Mirror* from 1932 and

---

58   Cf. Klonk, "Die phantasmagorische Welt," 134.

Fritz Winter's *Komposition vor Blau und Gelb* from 1955 facing each other in the central hall on the second floor of the Fridericianum (figs. 3 and 4).[59]

It should be noted here that some physical preparations were necessary to make this comparison possible. That meant not only to bring these artworks physically together in one room as it is always the case with art exhibitions. Bode commissioned the work by Winter, but it was eventually too broad to fit on the movable wall that Bode had in mind to confront the two paintings. Therefore, as we learned from a report of later restorers, Bode decided to scale the panel down by folding the canvas on all four sides to the rear.[60] The picture was literally and physically "trimmed" to make this comparison possible on the material level.

But what *tertium comparationis* is produced by this comparative confrontation of two paintings from different periods and different countries? Both are paintings but show decisive differences in format (Picasso's upright format vs. Winter's landscape format). There is, however, a similarity within the mode of referentiality: Both paintings do not follow a mimetic conception of referentiality but show the value of formal pictorial qualities within the logic of the image. In Picasso's *Girl before a Mirror* the artist even thematizes this question of referentiality in a non-mimetic mode with regard to the motif

---

59   Picasso, *Girl before a Mirror*, 1932, oil on canvas, 162.3 x 130.2 cm, Museum of Modern Art, New York, Gift of Mrs. Simon Guggenheim and Fritz Winter, *Komposition vor Blau und Gelb* (today entitled: *Durchbrechendes Rot*, 1955, oil on canvas, 381 x 618 cm, Museum Abtei Liesborn des Kreises Warendorf, Wadersloh-Liesborn (Leihgabe Fritz-Winter-Haus, Ahlen). See for the contrasting juxtaposition of these two paintings Klonk, "Die phantasmagorische Welt," 143. Grasskamp, documenta, on p. 21. Hemken, "Kuratorische Steuerung kultureller Diskurse," 134–135. Moser, "'Kunst [ist das], was bedeutende Künstler machen'," 40–41. Spies, Christian, "Fritz Winter. Kontinuität und Experiment." In *Fritz Winter. Vom Bauhaus zur Documenta*, edited by W. Utermann, Dortmund: Verlag Galerie Utermann, 2018, 34–47, esp. 34. Spies emphasizes Winter's specificity: "Fritz Winter was not just one German post-war painter among many. Like no other, he represented an art that, ten years after the end of the war, was redefining its regained freedom, not just in its coming to terms with the most recent past but also in terms of its status in the present" (37). Furthermore Winter's biography was highly marked through the National Socialists: "in 1937 his works in public collections were confiscated as part of the 'Degenerate Art' campaign and accompanied by an order forbidding him to paint and exhibit […]." (39/41).

60   See Herpers, Iris, "Fritz Winter, 'Durchbrechendes Rot'. Restaurierung und Transport eines großformatigen documenta I-Gemäldes," *VDR-Beiträge zur Erhaltung von Kunst- und Kulturgut* (1), 2011: 61–67, esp. 63. I thank Peter Gaida for hinting at these circumstances.

of the mirror (fig. 5).[61] At the same time, a gradual difference between the two paintings becomes obvious: While Picasso deals with a recognizable motif (and therefore offers some kind of mimetic reference), Winter's painting is non-objective (fig. 6).

Figure 5: Pablo Picasso, »Girl before a Mirror« (Bois-geloup, March 1932), 1932, oil on canvas, 162.3 x 130.2 cm, New York, Museum of Modern Art (MoMA), Gift of Mrs. Simon Guggenheim, Acc. n.: 2.1938.

© 2020. Digital image, The Museum of Modern Art, New York/Scala, Florence.
© Succession Picasso/VG Bild-Kunst, Bonn 2020.
Creative Commons license terms for re-use do not apply to this picture and further permission may be required from the right holder.

Through this curatorial practice of comparing, a continuous progress between Picasso's work of art from 1932, as representative of postwar art, and contemporary art is visually predicated. Moreover, the comparative viewing claims a continuous connection—and at least equality—between European

---

61   With regard to the motif of the mirror, see Meyer zu Eissen, Annette, *Spiegel und Raum in der bildenden Kunst der Gegenwart*, Bonn: Univ. Diss, 1980.

*Figure 6: Fritz Winter, »*Komposition vor Blau und Gelb« *(today entitled:* »Durchbrechendes Rot«)*, 1955, 381 x 618 cm, oil on canvas, Museum Abtei Lies-born des Kreises Warendorf, Wadersloh-Liesborn. Leihgabe Fritz-Winter-Haus, Ahlen*

© Fritz-Winter-Haus, Ahlen

modernism and German contemporary art—despite the break through the Nazi regime. The curatorial practice of comparing, staged here as a confronta-tion to show the progress of-non mimetic, abstract art, becomes even more obvious through the fact that the paintings hang on the left and right walls of the exhibition room.[62] Thus, the visitor is situated in-between the two or better: within the progress of abstract painting and the modernist historiog-raphy of art. As a bridge-builder, the visitor at the same time observes the progress and is an integral part of this European project of modernism, a position which was assumed as important within the young FRG. The histo-riography of postwar art history—as it was staged in the *documenta I*—played a decisive part in this political ideoligization.

---

62   For a closer examination of these artworks Hemken, "Kuratorische Steuerung kulturel-ler Diskurse," 134-135.

4. Conclusion

The historiography of German postwar art has often been told, even with regard to *documenta I*. In this context, two strategies have been emphasized within the research on this exhibition: the demarcation from National Socialism as well as from Social Realism.[63] Moreover, the role of *documenta I* within the narration of modernism has been the topic of many surveys: the discourse of the exhibition propagated a formalistic understanding of the image, claiming the progress of abstract, non-mimetic depiction.[64] On the basis of this research, the paper has taken a closer look on how this narration has been told, in the written survey of Haftmann's *Painting in the Twentieth Century*, on the one hand, and in the exhibition, on the other. In the comparison of the mode of narration the focus was on the ordering of time. It could be verified that the temporal structure of the narration of modern art history was achieved by practices of comparing, which produce continuity and discontinuity. The practices of comparing changed, of course, between written text and a staged exhibition. The different mediality, for example, required trimming of Winter's panel to make it suitable for a confrontation with Picasso's painting *Girl before a Mirror* on two movable walls. The temporal dimensions also changed. While Haftmann's survey *Painting in the Twentieth Century* produces a temporal order by means of comparing with regard to discontinuity, towards mimetic depiction in pre-Romanticist art and towards mimetic depiction of Socialist Realism, he also constitutes continuity with regard to prewar art and the European abstract art movement. The *tertium comparationis* was always the mode of referentiality. The narration of the progress of nonmimetic, abstract art, the story of modernism, was achieved through these practices of comparing and their production of continuity and discontinuity. In contrast, the *documenta I* exhibition relinquished the mode of discontinuity by neither preparing a comparison to paintings in the mimetic mode of depiction from pre-Romanticist times nor to the mimetic, figurative art of Socialist Realism.[65] On this basis, the exhibition staged the quality of continuity

---

63   Cf. Katja Hoffmann, who refers to Harald Kimpels research on *documenta*: Ausstellungen als Wissensordnungen, 93.

64   Cf. Hoffmann, Ausstellungen als Wissensordnung, 76–78.

65   There is one important exception with regard to the medium of photography: The entrance hall with its photographies showing historical works of art from different countries and epochs. Cf. Grasskamp, *Die unbewältigte Moderne. Kunst und Öffentlichkeit*, on p. 83.

all the more: through contrasting artworks from the pre-war era with ones from the postwar era as well as European artists with German artists—as was examined with the case of Picasso's *Girl before a Mirror* and Winter's *Komposition vor Blau und Gelb*. Practices of narrating and comparing play an important role within the construction of such politically important historiographic concepts. But to take them seriously as a subject matter of research also means to differentiate between the various modes of their production: in written texts and in staged exhibitions.

Authors

Martin Carrier, Department of Philosophy, Institute for Interdisciplinary Studies of Science, Bielefeld University

Joris Corin Heyder, Department of Visual History/Art History, Faculty of History, Philosophy and Theology, Bielefeld University

Oliver Hochadel, Institución Milá y Fontanals de Investigación en Humanidades, Consejo Superior de Investigaciones Científicas, Barcelona

Britta Hochkirchen, Collaborative Research Center SFB 1288 "Practices of Comparing," Department of Visual History/Art History, Faculty of History, Philosophy and Theology, Bielefeld University

Rebecca Mertens, Collaborative Research Center SFB 1288 "Practices of Comparing," Faculty of History and Philosophy and Theology, Bielefeld University

Christine Peters, Collaborative Research Center SFB 1288 "Practices of Comparing," Faculty of Linguistics and Literary Studies, Bielefeld University

Carsten Reinhardt, Department of History, Institute for Interdisciplinary Studies of Science, Bielefeld University

Hans-Jörg Rheinberger, Max Planck Institute for the History of Science, Berlin

M. Norton Wise, Department of History, UCLA

GPSR Authorized Representative: Easy Access System Europe, Mustamäe tee
50, 10621 Tallinn, Estonia, gpsr.requests@easproject.com